Kirtley Library
Columbia College
Missouri 65216

Marketing for the
Non-Marketing Executive

Marketing for the Non-Marketing Executive

Houston G. Elam
Norton Paley

 A Division of American Management Associations

*To Janet Elam
and to Annette and Bessie Paley*

Library of Congress Cataloging in Publication Data

Elam, Houston G
  Marketing for the non-marketing executive.

  Includes index.
  1. Marketing. I. Paley, Norton, joint author.
II. American Management Associations. III. Title.
HF5415.E49   658.8   78-1722
ISBN 0-8144-5465-8

© 1978 AMACOM

A division of American Management Associations, New York.
All rights reserved. Printed in the United States of America.

This publication may not be reproduced, stored in a retrieval system, or transmitted in whole or in part, in any form or by any means, electronic, mechanical, photocopying, recording, or otherwise, without the prior written permission of AMACOM, 135 West 50th Street, New York, N.Y. 10020.

Second Printing

Preface

How important is marketing? What do these characters really do? Is there really a theory behind their activities, or is it good guesswork? These are typical questions raised by serious-minded executives in many organizations.

Other, different questions: Why does marketing cost so much? How much of that promotion dollar is wasted? Why is the selling price so much higher than the factory price? How deceptive are marketing managers?

Marketing, and thus marketing managers, are vulnerable to such questions, for they are involved in very dynamic areas of the firm, are in the marketplace, and are involved in many of the activities that are at least partly visible to other managers, especially when something goes wrong.

At the same time, it is apparent that the marketing executive and the functions represented are becoming increasingly more central to the firm. The relationship to technology, market shifts, and increased competition from all sides makes the marketing role increasingly critical.

Under such circumstances, it is only logical that executives and aspiring executives in all areas of the organization would benefit from a basic understanding of the marketing manager's role. This puts the marketing manager in touch with the company on several levels—it provides an overview of the company as a whole, and points up the relationship between personal responsibilities and the marketing function.

The accountant would certainly benefit from knowing why it is that several of the smaller cost centers have more significance to marketing than might be true of the larger financial picture, and that those cost centers may have more potential as estimators than they may have as cost centers alone.

In the same way, research and development personnel might be surprised to learn that marketing executives use the same scientific approach they use in basic research, only here it is applied to the complex area of consumer behavior and market potential. The relationships among production, personnel, legal, logistical, and control staff are equally related in a variety of ways.

Keeping the non-marketing manager in mind, this book uses familiar terms and tools. The marketing manager is viewed as any other manager in the context of planning, organizing, analyzing, and making decisions within the context of his or her responsibilities. The flow of terms and concepts should be familiar and reassuring, even when the uses of data and the theory and objectives of the marketing manager are somewhat foreign. In keeping with the assumed managerial or premanagerial audience, certain basic terms and concepts are touched on. The entire text is built on the assumption that the reader already possesses an interest in the subject and is looking for the maximum number of facts and insights in the shortest reasonable amount of time.

Even managers should have an opportunity to review what they have learned if they want to. We have provided an appendix that includes both short-statement exercises, which we have called brainteasers, and more extensive statements called thought-provokers. This material can be used to test what you have learned from your reading. Many executives find these exercises to be useful indicators of their real achievement of knowledge.

As in all such projects, the authors received much assistance from many sources, many of them difficult to identify. The book gained considerably from the comments of three anonymous readers, and their courage, candor, and support are hereby warmly acknowledged. Mr. Eric Valentine, Acquisitions and Planning Editor, AMACOM Division of American Management Associations, encouraged us and followed through in an especially effective liaison. Mrs. Elizabeth Bailey played a most important role in seeing not only to the care and correction of the manuscript, but also, and at the same time, to the fact that other managerial responsibilities were continued.

David Fogg of the Montclair State College Audio-Visual Aids Department created the attractive illustrations in spite of the foggy perceptions of the authors.

A special tribute goes to Mary A. McGarry for many services performed effectively and with style.

And, of course, to our families, especially our wives, Janet Elam and Annette Paley, we acknowledge the hardship that this and similar projects have placed on them, and we are deeply appreciative of their support.

Houston G. Elam
Norton Paley

Contents

PART ONE / ENVIRONMENT

Chapter 1	Marketing . . . the Unique Function of the Business	3
Chapter 2	The Role of Marketing in an Environment of Change	11

PART TWO / PROCESS

Chapter 3	The Marketing Executive's Need for Planning	33
Chapter 4	Collecting the Data: The First Step	51
Chapter 5	Tools for Analyzing Marketing Data	86
Chapter 6	Developing a Marketing Plan	117

PART THREE / STRATEGY

Chapter 7	Orchestrating the Marketing Dynamics	153
Chapter 8	Activating the Marketing Strategy	239
Appendix	Brain-Teasers	241
	Thought-Provokers	249
Index		256

PART ONE
ENVIRONMENT

Chapter 1
Marketing...
The Unique Function
of the Business

Marketing is the distinguishing, the unique function of the business. . . . Any organization in which marketing is either absent or incidental is not a business and should never be run as if it were one.

–Peter F. Drucker*

These statements were penned more than a quarter of a century ago. They came at a time when nothing less than a major revolution was beginning to take place in executive headquarters buildings as the concepts of modern marketing management began to take shape.

Now that the revolution has taken place, the marketing executive has become fully entrenched within

---

* *The Practice of Management* (New York: Harper & Row, 1954), pp. 37–38.

the organizational hierarchy. The functions performed by these executives influence decisions in every area of the organization. Which products, promotions, styles, shapes, colors, smells, sizes, slogans to use all fit into the marketing executive's carefully planned approach. The marketing of every item is complex in its own right. Distribution of products represents a case in point. Consider how complex a task it is to distribute bottles of aspirin tablets to 40,000 outlets around the country and still come out ahead, even if they are sold at less than a dollar each.

How does the marketer know what the customer really wants? Is there a system to marketing, or does it really come down to luck? No area of a company is more vulnerable to such iffy assertions than marketing, and yet it works. Is it because there is a speical logic, theory, or process? The answer to these questions is *yes*. There are theories that apply, and there are sound and strong reasons for the approaches used. Knowledge of the facts and theories is invaluable to you, the non-marketing executive, in getting some perspective on the role of the marketer in your organization. Furthermore, it will complement the way you function in your own role, as both areas are mutually dependent.

THE MUTUAL STAKE IN SURVIVAL

Some time ago, a noted business consultant and philosopher, Wroe Alderson, pointed out that all managers have a "mutual stake in survival," that there is no effective way for a manager to advance continually at the expense of others, and that unless a manager understands the pressures, expectations, and basic re-

Part One   Environment   5

quirements of the jobs of others, he or she could be doing so inadvertently. Alderson's solution was simple: find out what the other manager's job is all about—what the pressures are, and what constraints are mitigating against that manager's success. For, if his or her function fails, and yours is tied in some way to it, you are better off knowing about it now than later. Alderson also felt there is every advantage to be gained by managers who communicate with one another about their functions and constraints, for in this way the entire system recognizes its mutual dependence.

Alderson's point seems almost too obvious. And yet, a look at a couple of questions concerning marketing and several selected other functions will show how well the pieces fit.

NON-MARKETERS QUESTION THE MARKETING ROLE

Some Production Management Questions
- Why do so many different products continue to be manufactured when some don't seem to sell well at all?
- Why are some of the best products eliminated from production, while other less efficient models go right on selling?
- Why does a company with a traditional strong suit continue to experiment and branch out into new product areas they fail in, only to try again in still another area?
- Why can't we establish a predictable relationship between price and cost?

- Why are some of the very best test products abandoned?

Some Financial Management Questions
- Why is so much money being spent on promotion and distribution—often in excess of 50 percent of a product's total costs?
- Why not use independent distribution systems instead of our own facilities to shift the payment dates forward and eliminate a substantial area of seemingly unnecessary costs?
- Why are there so many virtually identical products, and why do they sell for less if we make them and sell them under "private" labels?
- Why don't we schedule our production more evenly over the year and save a substantial amount of money?
- Why don't we use tighter credit policies?
- Why do so many marketing people have expense accounts?

Some Headquarters Staff Questions
- Why does the marketing group continue to use outside firms and consultants in such areas as advertising and marketing research when such services could be provided much more conveniently and much less expensively from within the firm?
- Why is there so much controversy concerning advertising and deceptive practices?
- Why are so many marketing decisions made so quickly, when a little more research and study would seem in order?
- Why does there appear to be so much debate and friction within some of those areas, and why do they stay on if they don't get along?

Part One   Environment 7

- Whose decision was it to advertise on that terrible TV show that nobody likes, instead of the top-rated show competing against it on the air?

Each of these questions probably has a good, strong, and defendable answer from the marketing standpoint, and each is important to the long-term interests of the firm.

INTERDEPENDENCE OF MANAGEMENT

Because management must work as a team, it is helpful to recognize and understand the point of view, as well as the needs and requirements, of the other executives within the management unit. One can thus begin to recognize why certain questions that seem so unimportant from your side of the table are given great weight on the other side. To the marketer, such issues as shortest possible delivery time and error-free high-quality products represent far more than maintaining standards—they mean a loss or gain of sales.

To the marketer, pricing is more than cost. It is a variable to be actively used and freely changed as part of a careful strategy for increasing profits. Sometimes a price position works as a fulcrum to deliberately bring about a short-term drop in sales or profits in order to accomplish a larger objective for future advantage. Each of these cases requires a marketing decision to complete the cycle, for without profitable sales from identified customers all is lost.

On the other hand, the marketing executive is at the mercy of a faulty production schedule, insufficient funding, or records that do not contain the information that will help identify potential markets or customers. Mutual dependence again.

MARKETING DEFINED

Marketing synthesizes the performance of consumer-satisfying business activities in the identification, coordination, and flow of goods and services from producer to user or consumer. Thus, the marketing executive must first focus on satisfying the consumer, and must orchestrate a wide variety of functions within and around the company in order to accomplish this objective.

The marketplace and the dynamics of competition are placed squarely within the term *marketing* itself and are carried over into its function. Furthermore, this broad definition of marketing implies that marketing executives do, and should, have a strong voice in what and how much of which items are produced on what schedule, and in a wide array of postproduction responsibilities. The implications of the marketing function are fundamental to the short-term and long-term operation of the business. This shift in business emphasis has made the marketing issue so important.

These comments should not be construed to suggest either that the marketing function is the only important business activity, or that it can, or should, take over other areas of the organization. All we are saying is that this particular function has so heavy an impact on the other functions that it merits special attention.

ORGANIZATION OF THE BOOK

A study of any complex subject matter requires some direction. Because the material discussed in this book is aimed at other, non-marketing, managers in industry, we use the general framework of planning and implementation, activities common to all managers in any specialization. The approach covers familiar territory, but special emphasis is given the needs and requirements of such subjects as data collection and analysis, policy setting, and the establishment of objectives, planning, implementation, and decision making, each from the perspective of the marketing manager or the marketing function.

In order to accomplish this type of presentation, the book has been organized into three sections—the marketing manager's environment, process, and strategy—each relatively independent of the others. By reading the chapters in their stated sequence, you will find yourself first placed smack in the middle of the marketing environment looking at the characteristics of the economy that have had a great impact on business in general. You will then proceed through many of the managerial tools and techniques of process and analysis with which you already have some familiarity. But you will be surprised at the particular character and essence of their use within the context of marketing. Your reading will then climax with the last section, which will give you an idea of the marketing manager's heavy involvement in the complex strategies that are part of that responsibility.

You may wish to skip ahead to look at the marketing executive's dynamic side and explore the types of strategic theory so essential to the function, before learning about the basic research and decision-making tools of the job. In that case, by all means begin with Part Three, Chapters 7 and 8, and then go back to Chapter 2. This approach could make the experience more dramatic and give you a better angle on the environment and process issues.

Whichever way you choose to look at the subject, once you are on the trail of activities and responsibilities, you will doubtless find yourself fascinated by the problems and opportunities, and have a clearer insight into why and how marketers do the things they do. You may discover new ways to improve total performance. At least you will know the right questions to ask in determining where things are going wrong.

Chapter 2
The Role of Marketing in an Environment of Change

Today's marketplace is both hectic and customer-dominated. The management of marketing activities within that marketplace has become an increasingly important job in a very competitive economy.

The American business system operates on the assumption that people—and therefore companies—perform best when they know that good performance will benefit them directly. For instance, if a company finds out that a competitor has introduced a new product similar to one it already has on the market, priced about the same as its product, but of superior quality, the company will have to consider ways of improving its own product. Otherwise, it will suffer a drop in sales and profits. When a company does improve its own product, the competition has the marketing disadvantage. It will then be the competing company's turn to search for a way to make its own product appear more valuable to the consumer.

Thus, the system exacts a continuing movement of product improvement within a market, a cycle of action and reaction to major changes in competition. And each move opens up the possibility of what Theodore Levitt calls the "expectation of benefits" for the consumer.

The stronger the competitive structure, the more demanding the pressure on the marketing executive. So successful have American marketing techniques been, that the last two decades have prompted a tremendous effort on the part of many socialistic and capitalistic countries to copy the marketing and distribution techniques of the more advanced capitalistic nations, such as the United States, West Germany, and Japan. Production has been tempered by sales, and sales have increased with style changes and various promotional strategies.

Coordinated marketing management as a formal activity emerged as the subject of serious study in the early 1950s, although a number of companies had already worked out their own systems and procedures before that. With such companies as General Electric, International Business Machines, General Motors, Eastman Kodak, and Procter & Gamble leading the way, formal marketing structures were gradually adopted by top management of most companies.

To make the most effective use of marketing, you have to understand the relationship between new technology and the pace of change in the marketplace, how the forces of environment and change affect each other, what part profit plays in planning, why the new concept in marketing has become so important, what the detailed benefits of planning are, and how to organize for efficient marketing.

TECHNOLOGY AND CHANGE

While change itself may be an unchanging characteristic of the American marketplace, the pace at which it occurs can and does change. The rate of change in marketing has been increasing steadily since the days of the Industrial Revolution. At first, it moved relatively slowly, but since World War II, it has been accelerating head over heels.

Every major change has resulted in a different consumer response in the marketplace, a different competitive situation. To be successful, a marketer has to anticipate and prepare for change, not just respond to it. Therefore, the more rapidly change takes place, the more vital the need for quick, accurate anticipation and adaptation.

There is a direct relationship between the production technology and the rate of change within an economy. As technology improves, the pace of change quickens. The development of modern production technology, and thus the story of American marketing, began with the Industrial Revolution of the late 1700s and early 1800s. The development of the technology continued at a somewhat uneven pace into the 1900s until industry met the severe economic crisis of the late twenties and early thirties. Technology then picked up again with ever greater speed and versatility with World War II.

The Growth of Technology
The Industrial Revolution grew out of two central ideas—that production could be improved if people were to work together in a central plant, and that workers should be encouraged to do specialized work,

for that would make training easier and output greater. These ideas sparked the development of central factories and of machines and methods that could be used in new, centralized plants.

With the introduction and development of the new industrial technology, production began to catch up, for the first time, with the demand for products. In addition, prices and wages became more parallel. There was more work that a man could do, more salary that he could earn doing this work, and more products for him to spend that salary on. The standard of living went up. The pace of change had begun to quicken.

The Civil War was the first important test of the new technology. Manufacturers met the test satisfactorily. The Civil War also encouraged the development of the still-young railroad system, and when peace reigned again, producers used the expanded railroad network to expand their sales over larger portions of the country.

By the end of the first quarter of the twentieth century marketers were handling a large number of products that provided a startling new standard of living. Although customers were given greater choice in the marketplace, the manufacturer and the marketer continued to dominate the scene. It was still a seller's world.

Switch in Market Emphasis
Depressions had come before, but the one that hit in 1929 and continued into the 1930s became known as the Great Depression. From a marketing point of view, the problem was not only one of trying to find ways to mend the economy as a whole, but also of each marketer's wooing those few and far-between customers who had any buying power left.

Production continued to be planned according to what manufacturers thought they could turn out most profitably, but more emphasis was placed on postproduction activities. Advertising and sales promotion were designed to influence customers to buy this product or that one. By the end of the Great Depression, promotion had become so important that four-color ads were commonplace in magazines, and radio commercials bubbled with jingles and snappy one-liners.

Increasing Customer Importance

The first rumblings of World War II exploded the last few lingering symptoms of the depression. With labor and materials being concentrated on the war effort, essential products and luxury products alike were in short supply. Innovation and substitution became a way of life during the war, making the early to mid-1940s what many consider the Second Industrial Revolution. New methods of food preparation and packaging opened up the old can of peas and introduced the new frozen product. Building materials included aluminum, glass, plastics, and concrete, replacing the iron, steel, and wood needed for war supplies. Among the many synthetic products that were developed, substitutes for rubber and butter took a permanent place in the postwar marketplace.

After World War II, the production lines reverted to consumer products. The business picture looked bright. People were anxious to see many consumer products that had been either very scarce or even impossible to find during the war. Consumers also had the money to spend. Those who had been in the armed forces and in the factories had saved much of what they had earned—having had so little available to spend it on during the war—and they wanted to go out and splurge.

To many manufacturers, planning seemed a simple matter again. All that most companies would have to do would be to pick out a good product, tool up for its production, turn the machines on, and simply let them manufacture that product for years and years. What they overlooked was how drastically the marketplace had changed. The mass production techniques picked up during the war had given manufacturers ideas on how to produce more products more quickly than ever before. They had learned how to innovate, substitute, and redesign with imagination and speed. The result: a marketplace flooded with new products and stronger competition than ever before.

Glutting the Market. Because business planning focused so strongly on the production line and all but ignored changes in customer demand, what was considered an almost endless potential market in the postwar forties became oversupplied in the early fifties. Here are three examples of slumps inverted by innovative thinking:

1. In 1946, automobile manufacturers began producing the first cars to come off the line since early 1942. People were eager for new cars. So successfully did the industry satisfy that hunger, however, that by the early 1950s there was a slump in sales. The market had been saturated. To counteract the trend, automobile manufacturers created a new line of personalized cars.

2. Before the war, iceboxes were far more common than refrigerators; automatic washers were a novelty; the home freezer was almost nonexistent. Yet

by the early 1950s, the market for these appliances was saturated to the point where justifiable sales could be achieved only through severe price-cutting. In fact, discounting was first introduced to encourage the sale of large appliances.

3. The home television market developed in the late 1940s. By 1955, the industry was in serious trouble—not because of a waning interest in television, but because practically everyone already owned a set by then. Assembly lines no longer had to turn out such large numbers of sets. How did they solve it? With miniaturized, portable sets.

Similar examples could be found in virtually every segment of every market and industry. The productive capacity of manufacturers had reached the stage at which almost any market demand could be met very quickly—only marketers had not yet learned how to plan effectively. The time between the introduction of a new product and the inevitable decline of its sales had dropped from many years to just a few years (today, in some markets, it is approaching a matter of a few months). Still, marketers did not know how to forecast and adapt to these changes. This time span is called the *product life cycle.*

No company can plan its fortune on the production of a specialized product with the expectation of rich profits for years to come. A simple example might be the company that produced the trim cardboard box that held four tomatoes, once sold by virtually every food store in the country. The manufacturer felt quite secure, for his product was protected by patents on the machinery necessary to produce the little cartons.

One day in the mid-1960s, a plastics equipment manufacturer looking for new markets for his machinery noticed the little cardboard boxes. He went to the tomato processors and told them that he had machinery that could turn out a plastic frame to do the job. The machinery would not only produce the frame, but would also wrap it and the tomatoes in a clear plastic seal. The plastic container would be stronger than the cardboard box, show the tomatoes clearly, and cost less than 25 percent of what the cardboard box was costing them. Within the next few months, cardboard box sales had dropped by 75 percent.

ENVIRONMENTAL FORCES

Not only does technology affect the changes within the marketplace, and thus affect the need for strategic marketing, but there also is an important relationship between market changes and the environment into which a marketer's product is introduced.

First, new products can change the environment they are introduced into by changing the way people live. Changes in the social environment are reflected in the marketplace in new attitudes about what people decide they need or want, and even where and when they decide to make their purchases.

Second, because of its particular characteristics, an environment—buying behavior, geography, response to promotion, competition, scarcity, acceptance by peer leaders—can either welcome or resist change and thereby accelerate or slow down the rate of change.

How Innovations Affect the Environment

New products and new ways of marketing them can have an impact on customer habits, both in the home and in the marketplace. Consider, for example, the introduction of the automatic washer. When the product first came on the market, housewives viewed it sidelong. Would it get clothes clean without shredding them? After a period of weighing possible problems against dirty piles of washloads, she welcomed the product. The new machine changed the housewife's life.

Before the housewife had an automatic washer installed in her home, she spent a very busy washday. She had to be at the washer most of the time—running the washer full of water, putting the clothes in, adding the soap, turning on the agitator. When the clothes had been agitated enough, she had to run them through a wringer into rinse water, change the rinse water at least once, and then run them through the wringer all over again to get them as dry as possible. None of these steps was automatic—she had to stand by through the entire procedure.

Compared to all that, the prospect of being freed to do other things while the wash was being done automatically firmly convinced the housewife to say goodbye to the old set of wringers. She was now in the market for new products, such as soap powder formulated specifically for use in automatic washers. And she was out of the market for other products, like a new set of wringers.

How the Environment Affects Innovation

Some societies, such as those of the United States and Canada, show little or no resistance to a new product or idea. A firm operating in such an environment must be constantly on the lookout for new products, lest a competitor get the jump on a new market. In a competitive environment, the pace of change is rapid, and planning must keep up with it.

But change isn't accepted so readily in all societies. Some view any change with hostility, seeing new products as something to be resisted. A considerable period of time can elapse before a new product is accepted as useful or practical. This may be the way of an entire society, or the reaction of a particularly conservative group within a society. To woo such innovation-resistant groups, the job of planning is to try to break down that resistance.

Acceptance of Change. Innovation-conscious environments accept change readily. Here newness alone can be a strong selling point for a product. A good example of how a plug for newness can affect planning comes across in the name of one of the better selling cleansers: Dutch Cleanser. When the product was first introduced, more than 40 years ago, it went under the name of Old Dutch Cleanser—Dutch because the Dutch people have a reputation for spic-and-span cleanliness, old to suggest that the product had the test of time behind it.

After World War II, several different types of cleansers were introduced and became popular. The new products were promoted as "stronger," "newer," and "faster." The general impression given was that they were better than the established products. It was not long before the manufacturer of Old Dutch Cleanser decided to meet the competition by altering the formula and changing the name. The company came up with New Old Dutch Cleanser. However absurd it might have seemed to talk about a new-old anything, the new product sold well, and the name change most likely helped.

By the mid-1950s, a still newer variation of the formula was introduced, along with still another variation in name. It was called New Dutch Cleanser, generally shortened to Dutch Cleanser. The "old" had vanished completely.

The American environment may be reaching the point where the term "old" will be considered a fault in a product. Few products stress their age in today's markets. A company may point out how long it has served its customers, but it is quick to emphasize how modern its processes and products are.

Resistance to Change. Resistance to change can be a total-environment reaction, or it can be a reaction of a group of people within a society. In the first case, it is usually rooted in the customs and mores of the society. In the latter case, it may be based on the beliefs or even the age of a particular group. For instance, many older people point out that today's products appear to be less well made than before, that they don't hold up as the earlier products did. What about planned obsolescence? We all know that some products are designed by their manufacturers to have an intentionally short life span.

This may have some truth but, as in so many subjects, the truth has many sides. Some products are intentionally made of less durable materials than they used to be, not out of any desire to cheat the public. For example, washpans used to be made out of galvanized tin or other metals; today they are made out of plastic. Perhaps plastic doesn't last as long as tin, but it's safer for whatever is being washed; plastic is considerably cheaper not only to manufacture, but to buy as well, and customers not only can buy it in many colors—they can buy a new pan whenever they decide to change the color scheme of their kitchens.

Most people recognize the difficulty in comparing yesterday's products with those of today. The environment has changed, and so have its products. What a marketer's planning is to assure is that the products offered will be acceptable in the environment being marketed in.

THE MARKETING CONCEPT

As a manager, you are fully aware of the need to provide for a strong profit. Profit tends to indicate the healthy mainstream of your activities. Costs are a major profit-retarding factor. It is only natural that you begin to focus on costs as the major factor affecting the delivery of such profit. Many companies have built their structures in a direct response system to this set of logical relationships. But it just may be that the very emphases being placed on costs and production constitute the biggest problem of all, for they tend to stifle those creative activities that might add to costs. New and different products offer the best hope for the future.

What, then, is the real generating force behind profit? Wouldn't it be selling products above cost? No doubt about it, but the point is far too important to be made so simply, as it tends to be overlooked. What counts is the *sale price*, not the *cost*. In other words, get the customer to prefer and be willing—even eager—to buy. This is where Henry Ford went wrong. He had an enormously efficient production line turning out his famous Model T. He is said to have offered the car in "any color you want, so long as you want black." The car was affordable and easily fixed at home. Why did he lose the market he had created? He opened up the field for others to develop cars that would compete with a variety of differing features, sizes, shapes, and colors. Almost all the new models cost more, and they made a lot more money. The approach to a successful future starts with the recognition that (1) change is inevitable, and (2) costs are by no means the only factors to consider. As we noted earlier, marketers were having quite a time keeping profits up just after World War II, not because there wasn't a strong demand for many of the new products, but because the production technology gained from the war effort satisfied that demand faster than they could keep up with. The problem further was compounded by the fact that product life cycles were getting shorter. Even if a company found a product with a strong market position, the firm could no longer count on a long, successful job run for continued profit.

Marketers had calculated profits in terms of how many units they would run off the production line. But by the early 1950s, no existing product or product assortment that was profitable one year could be guaranteed of doing well the next. The new technology created the need for a new system of planning for profit.

Production Orientation

In its simplest form, *production orientation* allowed a company to choose those new products it decided its production line could manufacture most efficiently and profitably. Back in the nineteenth century and the beginning of this century, this system could keep costs down because only those products that the company's machinery and labor force could produce with little need for retooling or retraining were added to a company's line.

The production-oriented system emphasized the

manufacture of products at their lowest cost (usually passing the savings along to the customer as a competitive weapon), making production experts the most influential managers in the company.

The approach seemed logical enough. What could be more sensible than making maximum use of existing facilities? It remained the logical solution for a long period of time, because the public did not yet have enough discretionary income to demand increased innovation or more rapid innovation on the market.

However, the production approach gradually crippled companies that sought to continue making profits in the changing marketing climate. As we know, profit is not related to production cost alone. It also depends on the marketability of the product. In our highly competitive marketplace, only those products that please customers win the sales—and the profits.

The production-oriented system saw profits dive from the 1940s and 1950s on because it neglected what should have been its prime concern: the customer. Not to say that costs are unimportant, for indeed they are. But costs are relative to opportunity. High production costs by themselves should not eliminate further consideration of a particular product—if there is a strong possibility that a particular product will return ample profits.

Market Orientation

Marketers came to realize that the days of planning profits in terms of production orientation were over. They were going to have to research what customers would want and then figure out how to give it to them. A new way to plan for profit: the marketing concept.

The *marketing concept* is geared toward the wants and needs of the consumer. A company that uses the marketing concept works out a profitable product development plan, systematically developing products within that area. It means that a company studies consumer habits and attitudes. It means that a company tests its ideas before putting them on the market. And it means that planning, in terms of both action and reaction to a market opportunity, must be a never-ending job.

The marketing concept can be used to develop a profitable product line through two channels: Either a new product can be developed, or a marketer's product diversification can be adjusted to meet market conditions.

New Products. New-product development involves choosing product ideas that appear to meet a customer need or want, analyzing its costs and revenues, developing it, testing it both in the laboratory and under actual market conditions, and, if the tests prove successful, introducing it to the marketplace. Careful planning gives the product a good chance of returning a respectable profit to the company. Table 1 compares the marketing orientation with production orientation.

Testing is the keynote. Crest toothpaste was tested again and again in market segments before it was brought to the general market. On the other hand, General Electric once tested a speaking letter device and then decided not to introduce it. The market tests indicated that the average potential customer saw little difference between the test product and the small cassette recorders already on the market.

The market orientation shapes the product development more than the production orientation

Table 1. Marketing vs production orientation.

BUSINESS FACTOR	COMPANY'S PROBABLE EMPHASIS	
	MARKETING	PRODUCTION
Corporate attitude	Focus on consumer priorities	Focus on company priorities
Research	Market-based	Technically based
Product development	Market-based	Technically based
Product mix	Ever changing line	Limited line
Packaging	To enhance sales	To provide product durability
Sales force	To sell and bring back information	To sell existing products

does. The market orientation requires a great deal of initial research. And in many cases, the research only shows that it would be better not to introduce the product. Still, those products that do pass all the tests have a far better chance of being just what the public needs or wants, and can produce strong, healthy profits for the firm.

Diversification. In addition to the systematic, planned development of new products, the marketing concept presupposes an open mind to all kinds of marketing opportunities. If tests show that an entry is likely to be profitable, a marketer may even consider diversifying his or her product assortment by entering a market in which the company has had no previous experience.

Perhaps the largest single problem faced by companies interested in diversification is that many of their executives have come up through the managerial ranks and have handled only the product of a particular market. They tend to be locked into thinking only in terms of that market. With this type of myopic vision, many possible opportunities are never noticed, simply because they do not seem to fit neatly into the company's pattern of activities to date.

For instance, the average railroad manager used to define his job in terms of running trains. He didn't see it as a means of providing space utility. The myopic "running-trains" approach made him resent the encroachment of buses, airplanes, and trucks. But once he saw the space-utility approach, he discovered that piggyback shipping—carrying truckloads of cargo on railroad cars and railroad carloads of cargo on ships—opened up all sorts of interesting possibilities for the kind of diversification that yields that important profit.

In the same way, some drug manufacturers spent so much time trying to find a particular drug to treat a particular symptom, they failed to see other possibilities for some of the drugs they developed. A drug that is ineffective on one market may prove perfect for another.

Mead Johnson's Metrecal is a good example. The product was originally intended as a baby food supplement, but had only moderate success in that market. However, Mead Johnson recognized that the same nutrients that made the product a baby food supplement would also be useful for adults who wanted to lose weight without losing needed nourishment. Mead Johnson began to market Metrecal in drugstores for dieters. Even then, success still eluded the firm until its marketing people realized that most people who try to diet by themselves purchase their requirements in a food store, not a drugstore. So the company took the product to the supermarkets, backed it with a strong promotional campaign, and Metrecal became a big success.

The marketing manager is surely one of the major focal points within an organization, and will remain there for some time to come. There is increasing evidence that new technology will continue to spin out completely new products and saturate entire markets at an even greater speed than before. Furthermore, marketing management is, and should be, held responsible for identifying, recognizing, and assisting the company in developing new products, and then seeing to their successful introduction and profitable sale in an ever more competitive atmosphere.

ORGANIZING FOR THE MARKETING CONCEPT

A generally accepted definition of the purpose of organization within the market-oriented company is to provide the rationale for bringing together the functions of sales, advertising, market research, logistics and storage, sales services, and product planning, and for creating new working relationships between marketing and the other business functions of production, finance, and personnel.

The dramatic nature of this reorganization can be seen in Figure 1. Note how functions that formerly operated under such areas as finance, engineering and research, and particularly under production, have shifted position into a new organization under the marketing banner. For example, forecasting; product planning, product services, and warehousing; and advertising were once separate groups in the production area; but as shown for the marketing orientation, become more integrated. Some services have been

Figure 1. The impact of the marketing concept on the corporate structure.

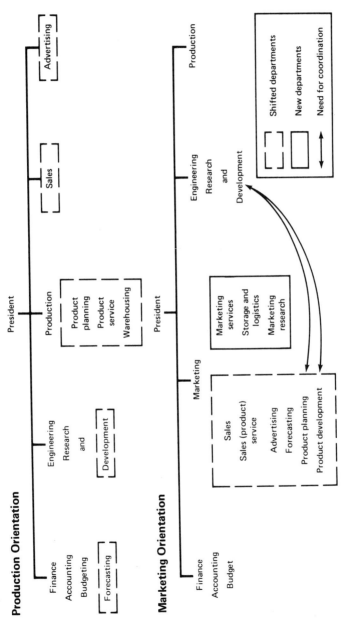

shifted and recast, with a new split personality (in two places—part of a department shifts, yet part of it stays where it was). For example, warehousing may be known as storage and logistics; development is known in some companies as production planning and product development. New functions, such as marketing research, marketing services, and product development from the consumer perspective, are created.

The resulting shifts redirect the focus of central management and of the corporate decision process. In this way, most companies have reshaped their attitudes toward their business objectives and have thrust themselves into many new ventures that would have seemed inconceivable only a short time ago.

PART TWO
PROCESS

Chapter 3

The Marketing Executive's Need for Planning

The various areas of the marketing function have a tendency to be volatile and complex. Marketing decisions affect especially large sums of money and resources. Such decisions tend to come with more rapidity than in other areas of the firm because of the particular relationship between marketing management and the rest of the firm on the one hand, and between this same management and the customer on the other. Under such circumstances, confusion and turmoil take over unless many complex elements are worked out ahead of time.

THE JOB THAT PLANNING DOES

Planning is essential to the success of any marketing operation. Without it, a company reacts to, rather than prepares for, the numerous situations it will confront over any given period of time. Without systematic planning, a firm survives and prospers on luck alone, never quite prepared for tomorrow.

Especially in the marketing area, planning helps a company shape its future success by measuring and analyzing every fact and by determining what it should do and how to do it. Working out objectives isn't easy, and it can be expensive, but the payoff more than makes up for the costs—and the risks.

Benefits of Planning

A marketer's resources are usually considered in terms of money, manpower, and materials. How well these resources are used determines the profitability of any marketing operation. Planning is what makes the difference. Only a good planning system can take into account the exact amount of resources available, the exact state of the market, and the probable trend in consumer wants and needs. Only then can the marketer resolve how to manipulate the resources in such a way that the company's offerings will meet market conditions, satisfy the customer, and have the best possible chance of producing a profit.

Planning can offer a company many benefits: organizing and facilitating clearer thinking, emphasizing alternative plans of action, coordinating and unifying efforts, facilitating controls, and reducing risks. This is especially true of the marketing operations, so let's spend a moment on some basic and seemingly obvious points and see if we don't still come up with an unexpected benefit.

Organizes and Facilitates Clear Thinking. Working out a plan, instead of simply reacting to changes when they occur, is a process that encourages us to think things through. We, as managers, are expected to consider all the facets and opportunities of a problem, work out goals and ways of reaching them, and be able to explain our particular choices. The planning process not only enables us to check the logic behind each move, it creates a forum for weighing and comparing the reasoning that goes into the plans other managers have come up with. They may make their own comparisons in a give-and-take planning session with other managers, or their ideas may be evaluated at other management levels. This process generally shows a continuity of thinking among managers. While each of us has our own problems and opportunities and our own way of handling or exploiting them, all have the company's goals in mind. This process of organizing ideas, approaches, and alternatives develops clear thinking, and ultimately the probability of good decisions.

Emphasizes Alternative Plans of Action. The more thorough the planning stage, the more opportunity management has to note, consider, and establish alternative courses of action. A marketer can never be absolutely sure of what the reaction to a move will be in the marketplace. By identifying planned alternatives, one for each possible reaction that may occur, the company will be prepared to keep pushing its marketing effort toward the goal, no matter what obstacles arise. The Mead Johnson example discussed in Chapter 2 demonstrates the value of recognizing that there is more than one way to reach a marketing goal. If Metrecal failed with infants, they would opt for adult weight-watchers (new market). If drugstores weren't the place to sell it, they would stock the supermarkets (alternative distribution).

Coordinates and Unifies Efforts. Planning assures that major undertakings will receive all the attention they need, so that every move is properly worked out and coordinated. The introduction of a new product is a complicated job. A product idea has to be generated, examined, and accepted. A design has to be worked out. Raw materials have to be located and a steady flow organized so that the materials are available when needed. A production schedule has to be set up. The price has to be determined. The distribution network has to be organized. A promotion campaign has to be worked out. And so forth.

Each of these efforts involves many problems; a weakness in one area can ruin the entire undertaking. Each and every job has to fit into a timetable. With proper planning, no job that should be done will be overlooked, and each will be completed according to schedule. Planning tells a company exactly what its need will be in one month, two months, six months, and provides a base for decision making that is consistent with its overall goal:

> 1. Quantity discounts can be taken on most of the products and supplies purchased, special sales by suppliers can be taken advantage of, and transportation of the goods can be arranged in the most economical way.
> 2. Labor costs can be kept as low as possible through careful scheduling of workloads and vacations.
> 3. Manufacturers can make bids on extra business if it appears the firm will not be running its production facilities at full capacity.
> 4. Retailers can reduce inventories for products whose sales are about to decline, and can increase inventories for products whose sales are about to increase.

Facilitates Controls. A careful plan is the only way a company can establish and maintain a system of good audits and controls so that management knows exactly how each activity within the company is proceeding. First, management charts the company's direction and sets goals that should be achieved by specific intervals. Next, management sets up a communications system that collects the data on actual versus planned results and provides feedback for each of those intervals. This system keeps management's finger on the pulse of what is actually being achieved, as compared to what had been planned. It can take the necessary corrective action if the audits and controls show that the results of one activity are way out of line.

Let's say a shoe store owner decides to stock professional ski boots. After studying the market conditions, the manager sets a goal of selling 40 pairs the first month, 50 pairs the second month, and 60 pairs the third month, reflecting the typical gradual fall to winter trends for this type of footgear. If the store sold 45 pairs the first month and had already sold 40 pairs during the first two weeks of the second month, the store owner would know that sales had been underestimated and that it would be wise to bring in additional stock. Without a control over the stock, the store owner might not realize he was going to run short until it was too late to get the additional stock. Sales he might have had would go to his competitors.

Reduction of Risk. The more time a manager has between making a decision and acting upon it, the higher the probability that it will be the right one, the profitable one. Decisions must be made carefully, and reviewing them or revising them according to last-minute changes in the company or the marketplace is essential to proper planning. Conversely, the later a decision is made, the greater the risk that it will be made in haste and without the proper information.

Action that has already taken place cannot be dismissed like a move in a sulky child's game of checkers. It can sometimes be changed by further action, but that means still more time and more expense spent on making the right move. In addition, when particularly sticky problems arise for which management finds itself poorly equipped to provide answers, there is still time to call in outside experts for their opinions. Without a cushion of time for planning, this option simply doesn't exist.

Managers need enough time to collect and analyze the necessary information that proper planning will give them, but they also need to move as quickly as circumstances allow, to give themselves a better chance of gaining or maintaining a competitive advantage. In Chapter 6, we go into this special time issue in greater detail.

To underline the importance of planning, here are some of the risks for a company that lacks a good planning system:

- *Too much reliance on luck:* A company that does not plan finds itself particularly dependent on the whims of fortune. If fortune is favorable, perhaps the marketer will thrive—for a while. If fortune is unfavorable, and luck has a habit of changing, the company will have serious problems.
- *Increased probablility of mistakes:* Improper planning or no planning at all increases the likelihood of the firm making a serious mistake. And a mistake can cut deeply into those important profits.
- *Poor morale and inefficiency:* A company that doesn't organize its planning has no guidelines for

top executives to distinguish between the effective manager and the sloppy manager. This makes for low morale among those managers the company needs most. The turnover of good personnel will be high, as those who know they can do a good job leave to seek more recognition elsewhere, and the company will have still more problems replacing them.

- *Narrowing of maneuverability:* Without planning, too often a company tends to commit too much of its resources, leaving too little in reserve. Such a company will find itself in a poor position to move rapidly if anything unexpected should happen—and in marketing the unexpected does happen. This point alone should be enough to encourage proper planning, for few things are remembered longer than an opportunity that might have been profitable but was lost because of lack of maneuverability.
- *Possible bankruptcy and liquidation:* The firm may not be able to pay its debts because it has overextended itself.

OBJECTIVES AND POLICIES

Marketing involves many more activities than some other functions. It covers not only the choosing of the products expected to have a good chance of success in the marketplace, but it also includes such activities as pricing, promoting, and distributing those products. Its overall goal is profit, but a marketing firm makes a profit only if it plans for the profit.

To do the job, a company needs clearly defined objectives and policies. These become the foundations upon which all its activities are built. For our purposes, let's agree that: *objectives* are the goals set by a company, either in short-range or long-range terms. *Policies* are the general rules by which the company operates, its guidelines for action.

Company Objectives
Objectives can be either long-term or short-term as determined by the time they are intended to cover. *Long-term objectives* range from periods of from two to five years, with near-term objectives being more specific than those intended for the long haul. These are sometimes called *basic objectives* because they are based on how management thinks both the company and the marketplace are going to change over a fairly long period of time.

Short-term objectives are projected for the next week, month, or year. They tend to be very specific in nature and frequently include target figures for sales, expense reduction, and the like. Because of their relatively precise nature, they are also known as *performance objectives*. Short-term objectives are more than modifications of previously stated long-term objectives. A considerable amount of reevaluation and reassessment goes into goal-setting for the short term, as it has to deal with continually changing situations both within and beyond the firm. A short-term study can mean a change of direction for a company's product or marketing emphasis.

Identifying Opportunities and Responsibilities. Any analysis of marketing objectives, particularly for short-range planning, identifies marketing opportunities. Sales quotas are set and particular competitive relationships tend to be defined. Members of management who meet or pass quotas and targets win the attention of the company for work well done. They are also expected to follow up on the success or failure of the established objectives. A company with well-defined, clearly assigned objectives is in a position to keep abreast of its progress toward those goals. To name just a few possibilities, marketing objectives deal with the share of the existing market held, or aimed at, by the company; penetration into new markets; the rate of sales growth; new products; methods of distribution; workforce needs; the competition; the markup of the product lines; costs; return on investment; and advertising.

Need for Establishing Specific, Written Objectives. In determining its marketing objectives, a company should be as specific as possible. To say "The company aims to increase sales significantly" doesn't give management a clearly defined goal to shoot at. On the other hand, if the stated objective of the company is "To aim to increase sales within the next 12 months by 15-18 percent over the average figures of the past three years while maintaining the same amount of net operating profit," then management can begin to plan in earnest. Examples of marketing objectives for a manufacturer and marketer of toys might include the following:

- To develop ideas and prototypes for new products that will make use of existing manufacturing facilities as well as help offset the seasonal fluctuations generally connected with the toy industry, stressing a middle-priced line of at least five adult games.
- To find new markets for the existing lower- and middle-priced game and doll lines.
- To increase the company's share of the market from 12 to 18 percent within 18 months in a market area now producing at least $500,000 gross sales.

- To explore ways of extending the product life cycle of the company's highest-priced toys without cutting their price and without spending more than 25 percent of the expected net profit increase.
- To achieve a return on investment of 20 percent on planned sales of $18 million.
- To restructure the organization of the company so that any variation between actual and estimated sales for each month is reported to management within three days of the end of the month.

There are two guidelines that can help a manager set up practical objectives for the company. First, although the business literature often talks about the advantage of long-range objectives covering a span of five years or more, it is usually better for a manager to start his or her campaign with goals that will be reached more quickly. Start thinking in terms of 6- to 18-month goals that will solve immediate problems or take advantage of immediate marketing opportunities. (The exception to this rule is in specialized fields of marketing in which commitments to capital investments must be made over longer periods of time.) Once these short-term objectives are achieved, the leap can be made to long-term objectives.

Some managers have an aversion to listing objectives. Their uncertainty stems from a basic inability to project their ideas very far into the future. But if they start out with short-term objectives and gradually progress to the long-term aims, their problem begins to take care of itself, as each long-range objective is based firmly on a series of short-range solutions.

Second, objectives should be geared to the company's resources. Ambitious objectives are certainly fine, and should be pursued whenever possible—as long as the company has ample resources to make such a move. But it should have good reason to believe that it has the resources to achieve whatever objective is set. For instance, if a company wants to break into a particularly competitive new market it examines whether a move in that direction is likely to use up too much of the company's resources. Obviously it would not want to win a share of the market, only to find that it could not supply the demand. Such an impractical objective would come under the general heading, "Don't bite off more than you can chew."

Company Policies
Objectives set the goals, policies determine general directions and procedures. Put another way, policies set the tone of the company as well as how it will attempt to go about its business.

Encouraging Initiative. Corporate and marketing policies should encourage new ideas and experimentation. To be effective, they must allow personnel to make mistakes without losing the opportunity to learn from them. The company must express willingness to permit its managers to retain their status, at least in the short run, while they are experimenting with new ideas, for it is those very managers who represent the lifeblood of the organization.

Assigning Responsibilities. Members within the corporate structure operate best when they know where the authority and responsibility lie for certain decisions. Policies would encourage creativity, but should also indicate who is to make which decisions and whom to go to for the answers to specific types of questions and for approval to act upon them.

Various Types of Policies. Like objectives, policies are established for various levels and functions of the organization. Policies at the corporate level are established in the areas of general conduct, personnel, physical facilities standards, and the like. Such policies tend to be broad.

It may be the general policy to produce the highest-quality product, regardless of the cost. It might also be extended to provide service to the customer in the event that the product fails within a stated period of time. This type of policy must be conveyed at all levels of the organization, by president and salesman, by production manager and warehouse clerk. Thus a manager's marketing activities are not separate and distinct from the company's policies. It reflects company policy through interaction with the other operating divisions. Any gap between policies and activities breaks the threads of the relationships that hold the company together, and results in a haphazard group of marketing activities without a solid purpose.

In the 1970s, policies began to broaden in scope and to take on a social responsibility unparalleled in business history. Organizations are becoming increasingly involved in the problems of society. While there are still many business operations that simply open the store, sell what is on the shelves, pay the workers a going wage, and then close the store at night, the tide is changing: To grow, a business has to live in both today's and tomorrow's environment, and its policies must reflect that environment.

Consumerism began to pack its punches in the late 1960s in order to give the buyer a fair deal. Concern with pollution, poverty, unemployment, crime, narcotics, and discrimination against minority groups has become an important issue in business policy. Business policies can reach out to meet new ideals, new standards, by:

- Maintaining high standards of product quality.
- Providing maximum benefits to the public by determining what the public wants or needs.
- Creating an atmosphere that encourages personal dignity among company employees; providing opportunities for advancement; paying fair wages; training the handicapped and underprivileged for better jobs.
- Selling products only at prices that will produce a fair and honest rate of return in the form of profit.
- Using honest, informative promotion to tell the public about products.

A company may decide to increase sales within a specific area, within a specific amount of time, which might require hiring additional salesmen. In this case, company policy would have to specify that there be no discrimination in hiring, that those chosen be properly trained, that they be offered fair compensation for their work, and that the new employees be treated as courteously and helpfully as possible by all company personnel with whom they come in contact.

A BRIEF LOOK AT THE CHAIN OF COMMAND

While our attention is focused on the values of planning and setting policies and objectives within the organizational structure, a word is in order on the chain of command. In the process of growing and expanding the organizational structure shifts according to new sets of relationships. Every time a new office is established, the manager must report to a superior, who in turn must supervise the department. The longer the chain of command becomes, the more difficult it is to be responsive to change, to retain the spirit of the team unit. Uniquely enough, the marketing concept provides a focus to restudy relationships, and the systems analysis people are of considerable value in reassessing those relationships within the structure.

Reevaluating the Existing Chain
In today's business environment, a manager has to react quickly, to maneuver and concentrate his or her marketing forces to gain the most profit in the shortest time span. That is the way to gain and exploit the marketing advantage. The proverb that "opportunities are fleeting" has added impact in today's technology-oriented, speed-conscious, highly competitive marketplace.

From a practical standpoint, the job of running the organization itself cuts into the time a manager has for planning a marketing move. Like keeping a house in order, there are always chores to be done, and these take valuable time. In addition, if the organization involves a long chain of command, even more time can disappear while information works its way up the chain and orders travel down.

How the Chain Grows. In a small organization, the president and his or her top aides are at the front of the forces, where the action is. The president sets objectives and establishes policies. Here, the organizational structure is a simple one. Information and directives are neither slowed down nor misinterpreted, since the long, cumbersome chain of command is nonexistent.

Our attention is on large firms, however, where the emphasis is on an increasing number and diversification of products, decentralization, and management by a group rather than an owner-manager. As the company grows, new organizational problems, people, job titles, and levels of authority enter the picture. With each new link added to the chain of command comes the increased possibility of turning the company into a cumbersome, inflexible organization where swift, coordinated planning can falter.

Salesmen in the field often feel something is jamming the home office mechanism they need to help develop sales volume. Missed opportunities are all too common simply because the "go" signal somehow got held up along the way. Even in the home office, managers who aren't directly affected often think there are too many people on the general staff and on the service staff and not enough directly involved in sales-producing work. In a burgeoning organization, it is true that home office staffs tend to grow faster than field staffs, yet the shortage of efficient managers remains a source of constant complaint. Of course, once a company realizes that such problems exist, efforts are made to regroup and reorganize the staff, which merely results in upsetting the efficiency of the organization without producing any compensating gains.

Cutting the Chain. The major goal of a company is profit. Profit is most often won by holding a market share in the face of competition or by gaining a share of a new market. One way is to come up with new or improved products, new enough or improved enough to defeat the competition. Another move is to develop faster, more efficient means of collecting and analyzing market information. Both these courses of action, however, involve long and careful research, experimentation, and testing, before they become operational. Research is essential. A company's honesty and reliability should be old-fashioned, but not its products or its systems.

A simpler, shorter way to improve and plan successfully for profit, not to be overlooked by any manager, is to simplify the system of control and to shorten the chain of command. This puts the company in a more flexible, maneuverable position. Every extra link in the organizational chain means lost time in getting vital information to and from the head office. Each unnecessary link weakens management's grasp of the situation and its possible solutions or opportunities because it becomes more remote from that situation. It not only means a multiplication of office personnel, but often takes the most efficient and productive of the sales force out of the field and brings them back to the home office to fill the new staff jobs.

When a company has only a few intermediate links or levels, its planning tends to be more dynamic and flexible, provided that each manager is assigned a manageable workload. Obviously, a flexible organization can achieve greater market penetration than a cumbersome organization can. It has the capacity to adjust quickly to varying circumstances and to concentrate on the moves most likely to achieve success.

Just how short a chain of command should be varies with the type of operation. Suppose a marketing manager has two subordinate middle-management levels between her office and the sales force. True, those middle managers take much of the detail work off her shoulders and she can concentrate on top management matters, but it probably takes more time for information from the sales force to reach her and more time for her directives to reach them. Furthermore, by avoiding direct contact with subordinates, she loses out on the intangible impressions that can herald problems or lost opportunities.

Now, suppose she eliminates one of the subordinate management levels, concentrating all the subordinate management authority and responsibility at a single mid-management level. What happens then? The marketing manager has shortened her chain of command and has gained the flexibility to react to market opportunities.

Mid-management Authority. That marketing manager did one additional thing when she shortened her chain of command: She increased the responsibility and the authority of the remaining mid-management executives. This has its advantages and disadvantages.

As recently as 20 years ago, large businesses still operated as single-command companies. Only the head of the company had the responsibility and authority to handle overall planning. Subordinate management, regardless of level, could do little more than make suggestions, when specifically asked for them, and carry out orders.

With the acceptance of the marketing concept in the 1950s, the emphasis on planning shifted to meeting customer wants and needs. Increasing decentralization of companies into flexible divisions or profit centers gave mid-management executives greater responsibility for planning. In fact, these mid-management executives saw the need for the change and brought about its acceptance, although it was granted begrudgingly by many senior managers unwilling to consider anything so radical as giving up or delegating authority. Mid-management executives became closer to market situations and better equipped to identify opportunities and risks more rapidly and in more detail. They used this proximity to produce more profits for the company while increasing their own managerial standards.

There seems to be only one disadvantage to mid-management authority. Increased responsibility within a specific field tends to give middle management an increasingly parochial view toward its own field. Managers are intent on defending their own units and are usually opposed to channeling resources into other units. This tends to divide the marketing effort fairly evenly throughout the company's operations and effectively prevents the company from concentrating conceivably unusual marketing efforts on a particularly attractive marketing opportunity.

However, today, more than ever, business must expect the unexpected and be able to adapt quickly and smoothly. Marketing success now involves taking advantage of fast-breaking local marketing opportunities, usually handled best by the mid-management executives. In addition, the increased responsibilities of middle management tend to make it a more flexible group, more useful to the organization as a whole. Long training and excessive time spent in one job may make a manager an expert in its execution, but such expertise is often gained at the expense of originality and flexibility, the two essential weapons needed to do today's planning.

A Possible Trend. One additional method that caught fire in the late 1960s and 1970s was the formation of venture groups by top management. The group acts as an independent subsidiary that reports only to the executive vice-president or president. It is given an operating budget to develop its own product and to distribute, price, and promote it as it sees fit. It can call on the parent company for research help, but otherwise the group is on its own. This helps eliminate "infighting" for funds.

Chapter 4
Collecting the Data: The First Step

What is the most profitable market for our products? What are our competitors doing? What do our customers really think about our products? What types of new products are customers looking for? Questions, important questions. The answers can go a long way in determining what success the firm will have in the marketplace.

Marketers, like other managers, need information. Although managers in most other areas of the firm use internal data for their decisions, marketers have to rely on data gathered outside the company on the customers and their desires, economic conditions, shifts in consumer lifestyles, and the foundations for sound marketing decisions and good marketing planning.

Collecting this data is part of the job of market research. Simply defined, *market research* is the application of scientific methods to the solution of marketing problems. The American Marketing Association's definition is more detailed: "Market research is the systematic gathering, recording, and analyzing of data about problems relating to the marketing of goods and services."

The definitions are very similar. One simply amplifies the other. The emphasis in one definition is on the word "scientific," and in the other on the word "systematic." Both adjectives underscore the importance of eliminating as much guesswork as possible from the decision-making process. Planning must be based on solid, pertinent data, collected and analyzed both scientifically and systematically.

Market research as a formal science or a set of activities is relatively new. Any marketer interested in early marketing information finds that surprisingly little organized marketing data was collected before the 1920s—and data is the heart of market research. After the turn of the century, marketing began to be considered enough of a subject in its own right to be added to school curricula. In 1917, U.S. Rubber and Swift and Company organized what were probably among the first company market research departments in this country. The first authoritative textbooks devoted entirely to marketing appeared in the 1920s. The first comprehensive collection of data on marketing in the United States was collected by the Department of Commerce, through its Census of Distribution, in 1929.

Today, market research is an accepted part of industry. Many companies have been formed for the sole purpose of carrying out market research on an independent basis, and most of the larger companies have market research departments of their own. However, a marketing company needn't be large to become involved with market research. Its techniques and methods can be used profitably by any marketer, and they must be used if a marketer wants to plan profitably.

Part Two   Process 53

By definition, market research involves handling the data needed for market planning. This process can be divided into two phases: collecting the data and analyzing the data. The first stage is the subject of this chapter, the second is discussed in Chapter 5.

Regardless of company size or whether data is being collected on a regular basis or for a single special use, to be handled by the marketers themselves, by a department within the company, or by an outside organization, the basic facts of proper data collection remain the same. Success in collecting data depends on knowing these facts—the types of data, the sources, the methods of collection, the sequential steps to follow, and the kinds of data collection systems used by companies today.

TYPES OF DATA

Every marketing operation uses a large variety of marketing information: facts about the marketing mix itself (the product, its price, its promotion, and its distribution); the company (its assets and liabilities, its reputation, its plans); competing companies and competing products; trends in the market and in the general economy; and customer habits, attitudes, moods, and behavior. There is virtually no end to what can be collected for market research. An important part of the job of collecting facts is to sift through what is needed for the task at hand. This is done by separating the facts into two groups.

Data can be either secondary or primary. *Secondary data* is information someone else has collected for another purpose, but which may come in handy for the task at hand. *Primary data* is original information collected for a particular project. These terms refer not to the value or usefulness of the data, but to its source.

Secondary and primary data can be either internal or external. *Internal data* is collected from within the company itself. *External data* is collected from sources outside the company. Most market research involves, and most planning requires, both types of data. Each type of data has its own characteristics, its advantages and disadvantages.

Secondary Data

A baby food company executive studying the new birth-rate statistics released regularly by the Bureau of the Census uses secondary data. The information collected by the Census Bureau for the federal government and its usefulness and interest to the manufacturers is secondary to the original reason for collecting it. But manufacturers need this information. Nowhere else could they obtain such detailed figures about their particular group of customers unless, of course, they were to set up their own private census units, an investment of time and cash that would put each out of business almost immediately.

Secondary information has three general advantages over primary information: (1) it is easier to obtain, (2) it is less expensive to obtain, and (3) some facts are available only in the form of secondary data (for even if a manufacturer did set up his own private census, he would have no right to demand that people answer his questions, a right that the federal government does have).

The major disadvantage to secondary data is that the user has no control over the original collection of the data and does not know how accurately, scientifically, or honestly the collection may have been made.

Some organizations that collect and distribute data do so for a promotional purpose or for a motive that may not be obvious—like supplying potential advertisers with fudged statistics on the readership of a magazine.

This is not to imply that the average organization deliberately distorts or changes figures, but that it simply organizes and displays the figures in its own best interest. Different inferences can be drawn from a single set of figures, depending on how the figures are listed and analyzed. Most sources of basic secondary data, however, are honest and unbiased. They are very willing to explain how the data was collected, what sources and methods were used, how the figures were recorded, and how they were analyzed. With this information, researchers can then decide whether or not the data is valuable for their purpose.

Primary Data

Primary data has two important advantages: (1) the precise facts needed for a specific decision can be obtained, and (2) collections of data not available from any other source can be built up. For instance, the owner of a small dress shop in a suburban town might find it difficult to base her choice of inventory on secondary data. What the average customer in the average store wants in skirt lengths may not match the taste of her own select group of customers. Instead of relying on secondary data, that marketer would be in a much better position if she were to collect primary data, probably by conducting a survey of her customers. She would then be in a better position to know what kind of stock her customers will probably buy.

The major disadvantage of primary data collection is that it usually involves a considerable investment of time, money, personnel, and facilities. Whether the results are worth that investment is an important decision a marketer has to make. Take that dress shop owner. She may have to hire a trained interviewer to question everyone who comes into the store for a two-week period. The project would involve not only the cost of the interviewer and then tabulating and analyzing the data collected, but the inevitable slight slow-down of business as customers sit down and chat instead of concentrating on making purchases. As long as the data collected and the decisions involved concern the major profit-producer of the shop (the inventory of dresses), it can probably be chalked up as a worthwhile investment. It would be a waste of that investment to spend that much money and time making a survey of what customers think about belts, which represent a very small portion of the shop's inventory.

SOURCES OF SECONDARY DATA

A researcher usually begins the search for suitable data by considering whatever data is available within the company itself. Internal data tends to be inexpensive and easy to locate, and the facts have the obvious advantage of already being geared to the company itself. Next, the researcher examines the various sources of external data. Obtaining secondary data from external sources is not necessarily more expensive than using internal sources, but it usually is. External data can be more time-consuming to locate and, unless the marketer hires a special firm to collect specific data, what is collected often contains considerable superfluous material. That means the marketer has to spend time sifting through a great many facts and figures before finding the ones that will be useful. Nevertheless, it still is the cheapest and quickest way, and sometimes the only way, to obtain the necessary information.

Internal Sources

Ideally, everyone within a company is working toward a common goal. Any information generated within a company should be freely available to and exchanged by any unit or department within the organization, regardless of its size. Unfortunately, the ideal is seldom realized. In some large companies, departments tend to be parochial about their own collections of data, preferring to retain them for their own use. In large and small companies alike, people lose track of data, when there is no central data-collection system. We go into this type of system in the last section of this chapter.

The first place a marketer should look for data is the accounting department, whether it consists of a large staff and a building full of records, or a ledger and a file drawer. The next stop is the sales department. The third possibility is to check the files of whatever divisions or departments do any work related to a particular marketing problem or opportunity. The major sources of readily available data within a company are:

- Financial statements, including the balance sheet and the profit-and-loss statement.
- Sales records for customer, product line, product, size of order, and so forth.
- The accounts-receivable ledger or customer ledger.
- The file of customer orders.
- Sales compensation records, including salaries and commissions.
- Sales expense reports, including expenses of traveling, hotel, meals, and entertainment.
- Sales reports (daily, weekly, and summary).

The information in these standard accounting and marketing records may be analyzed according to product, customer, area, or whatever frame of reference the marketer has in mind. For instance, a manager collecting facts about the profitability of a particular product would find useful information in just about every one of these records. The profit-and-loss statement would show the ratio of profits made by this product to those made by all products combined. The file of customer orders might indicate whether supply of the product was keeping up with customer demand. Sales compensation records would indicate whether the salesmen handling the product were having problems moving it, while expense records would show how much selling investment it took to sell the product. The sales records themselves would show how the product was coming along within its probable life cycle.

In addition to the information provided by the basic records, the company has price lists, customer correspondence, and service records. Many companies have built up collections of data used in other areas of planning that might be useful to the manager in any area of market research. A researcher in a large company can spend a long time exploring company sources of information before running out of material. Even in small companies, normal records contain far more information than many marketers realize.

External Sources

A manager can learn much from the company's own files, but external sources are more likely to know the trends in the marketplace and in customer preferences. Investigating these sources can involve considerable time, as when a researcher hunts through library shelves seeking the information needed to piece together a collection of facts. It can also be as simple as writing to an organization and requesting a specific list, chart, or brochure. The most useful outside sources of secondary information include libraries, advertising agencies and media, market research agencies, trade associations, universities and foundations, and the government.

Libraries. A tremendous amount of secondary data in the literature is carried by libraries of all kinds, public and private alike. The smaller marketer will find that the local public library has a surprising amount of useful information on its reference shelves. The larger marketer may want to explore the collections of libraries in universities, institutions, and even large companies in the field.

Public libraries are open to all. University, institution, and company libraries are usually open to anyone who can give a logical reason for wanting to examine their material. Librarians are usually available to give some help in locating information, but the market researcher will have to do much of the work himself or herself.

Source guides to help track down specific books and magazines for pertinent data include:

- *The library's card catalog:* Lists all the books and periodicals contained in the library by subject, title, and author.
- *Business Periodicals Index:* A cumulative listing of all articles appearing in the business press, by subject.
- *Monthly Catalog:* A listing of all publications put out by the federal government during the preceding month, arranged alphabetically by name of the issuing bureau.

- *Marketing Information Guide:* A summary, issued monthly, of the major marketing and distribution books, reports, and articles.
- *New York Times Index:* Synopses of articles in that newspaper, with reference to date and location.
- *Wall Street Journal Index:* Either the headline of or a synopsis of each of that newspaper's articles.

There are also a number of books published and updated regularly that contain information about where to find what facts. A number of volumes of condensed and summary statistics are published regularly in pamphlet or book form, primarily by the federal government. Although librarians seldom have the time to locate the exact information a marketer needs, they are always willing to guide marketers to the books that will help them find their way to the right material.

Advertising Agencies and Media. Most advertising agencies and the media they serve (such as newspapers, magazines, television and radio stations, and networks) have set up their own research departments to serve marketers interested in using advertising in the media. Most of their research effort is concentrated on the customers reached in the particular medium and the effectiveness of certain types of advertising.

Agency research tends to be quite specialized, and an agency will sometimes undertake a special project for a marketer. Media research is often more promotional than professional, its goal being to promote the media as the right place for the marketer's campaign. Although reputable research departments do not distort the results of their research, it is wise to remember that the facts most readily available from such sources are those that encourage advertising in the media. However, this type of research can produce many interesting facts about a given industry that a marketer can use.

The results of media research are usually handed our freely, often published as giveaway information by the medium itself. Advertising agencies absorb some of the costs of obtaining data in the agency's regular commission on the sale of advertising, reserving a fee for any special research work they agree to perform.

Market Research Agencies. Market research has become such an important activity in its own right, a number of companies were set up just to specialize in it. Market research work is done by two types of company: consulting firms and syndicated data firms. They and any other companies that engage in market research make use of market research service firms.

Consulting firms in the market research field specialize in solving specific marketing problems brought to them by clients. They work for a client as independent contractors who do a specified job for a specified fee. Most consulting firms employ relatively small staffs of trained specialists, and often handle work only within a specific industry or field. Anyone knowledgeable in a particular industry can recommend reputable consulting market research firms in that industry; local banks and business bureaus also suggest names of local firms.

Syndicated data firms specialize in collecting certain types of information and then selling them on a subscription basis. Thus, a syndicated data firm does research for others in a general way, but it does not undertake a specific problem for a specific company. Such a firm usually concentrates on a particular industry or a particular kind of information. Typical specialties include surveys of brand recognition, public opinion, fashion trends, advertising volume, and traffic counts. Among the larger and better known syndicated data firms are A. C. Nielsen, Market Research Corporation of America, Daniel Starch and Staff, Gallup and Robinson, the F. W. Dodge Corporation, and R. L. Polk Company.

The market research services company neither collects data nor conducts any facet of basic research, yet it provides a service for those that do. Most firms specialize in one form of service, and such specialties include providing trained interviewers, tabulating personnel and equipment, and even printed and bound research reports.

Trade Associations. Every trade association gradually builds up a fund of information about the industry or field it serves. Seldom does a trade association undertake research for an individual member. Most of the data that a trade association does collect are available to member companies.

Most trade associations collect periodic reports on members' operating figures. Although the figures each company supplies are usually kept confidential, they are used to work out statistical tables that show the typical, average, or median figures reported. These statistics are useful in gauging the operating trends among members, and if the association is important enough in its industry or field, the figures obtained reflect the industry or field as a whole.

The advantage of trade association figures is that they tend to be the only unbiased figures available for the industry as a whole. Yet, because such organizations are not primarily set up as research organizations, the data is neither collected nor analyzed in the most scientific way. Furthermore, the average figures may be based on a very small sampling of the industry.

Universities and Foundations. Most state universities and many private universities now have business research bureaus or departments that do basic work in collecting useful business information. These bureaus or departments seldom undertake a project for a specific company unless that project is one whose results will benefit the industry or the business world as a whole. Here, the subject of the research is determined by the university, the work is handled by the university, and then the results are made available in published form to anyone who is interested.

In addition to groups within the universities, there are also a number of research organizations affiliated with universities that conduct basic marketing research. The better known of these are the Stanford Research Institute, the National Opinion Research Center of the University of Chicago, and the Survey Research Center of the University of Michigan. Again, results usually appear in published form, either as an article or book.

Foundations also do considerable research. Those foundations most likely to be involved in marketing research and thus have marketing data available include the Brookings Institution, the National Industrial Conference Board, the National Bureau for Economic Research, and the Twentieth Century Fund.

Most of the publications put out by universities and foundations alike are reviewed or listed, upon publication, in one of the guides discussed earlier in the section on libraries. In fact, the local public library is one of the best places for an inexperienced small marketer to begin a secondary data collection, for the library usually houses a fairly wide selection of material from all kinds of sources.

Federal Government. The largest single source of secondary marketing data is the federal government. The surveys made by the Bureau of the Census provide the foundation of the entire structure of marketing knowledge in this country, and these are supplemented by research done by other bureaus within the federal government, such as the Bureau of Labor Statistics, the Federal Power Commission, and the Bureau of Agricultural Economics.

In addition to simply collecting data, some federal agencies do other types of research as well. For instance, what regulations or standards are needed in a particular industry or field? The by-product of the research may be an interesting collection of new data. The Federal Trade Commission, the Food and Drug Administration, and the Federal Communications Commission have carried out elaborate research whose results are available to anyone.

The vast collections of data produced by the federal government are available in printed form for remarkably low prices. The local library can give a marketer information about what is available, but those who live in areas that have a regional office of the Department of Commerce, which is the parent of the Bureau of the Census, should investigate the services of that office. Those offices with well-trained staffs and good supplies of material can produce quite an array of data in a remarkably short time.

State and Local Government. Agencies of the state and local government abound with information that may prove helpful. Although such data varies both in quality and quantity by state and local governing agencies, where it is available and accurate, it is particularly useful because it is more local in nature. The more local or regional it is, the more probable it is to reflect important trends in the area of most interest to a regional company.

At the state level, valuable data can be obtained from the license bureaus of various types from automobile to liquor authority boards: construction figures, state income and property tax figures, population and industry-reported figures, state employee salary figures, to mention but a few. Each state issues either an almanac or state of the state message handbook, which contains large amounts of current or budgeted data. The larger industrial states, such as New York, California, Illinois, and others, provide rather substantial amounts of information collected carefully and thoroughly. Local governments are a good source for data on home starts, utility connections, tax-rate data on homes and businesses, real estate transactions, registration fees for business operations, and local tax assessments.

METHODS OF COLLECTING PRIMARY DATA

Primary data is original information collected for a specific purpose. The marketer may handle the research job himself or with his own staff, or he may hire a research firm to do the work for him. Regardless of who does the work, the project is tailored to the specific data needs of the marketer. The kind of information the marketer wants is collected by the methods chosen by the marketer.

There are three basic methods used to collect primary data: the survey method, the observation method, and the experimentation method. Each has its own advantages and disadvantages. The one to use depends on such factors as the kind of information needed, the accuracy and objectivity required, and the amount of time, money, personnel, and facilities available.

Survey
Contacting people for specific kinds of information is the survey method of collecting data. The contact is usually made in person, over the telephone, and through the mail. The people contacted may be everyone in a particular group, such as all the members of a trade association, or they may represent a sampling of a group, such as every twentieth name in the local telephone directory. The kinds of information sought may be hard facts or opinions.

The survey is the most widely used method of collecting primary data. The value of the data depends on how the survey is organized and carried out. Every aspect of a survey must be carefully planned and executed if the results are to be worth the time and money spent.

Surveys can be done by the smallest marketer, yet they are a popular tool of some of the largest marketing companies. A specialty shop owner may stand in the aisle of his store several hours a day for two or three days, asking customers five or six questions and then mentally analyzing their answers to help him pinpoint any changes in what his customers may want. This same kind of information might be collected by a large store that sends out a questionnaire to every charge account customer on the store's list.

A survey can cost very little or run into a lot of money. A personal interview, for instance, can cost anywhere from around $2 to $500 per person interviewed, depending on the qualifications of the interviewer, the travel costs involved, the type of person being interviewed, and the length of the interview.

The Form to Use. The most common survey forms are the interview and the questionnaire. The *interview* is a survey conducted by one person of another person, either face to face or over the telephone. The *questionnaire* is a printed survey sent to a person; it may be mailed or handed to the person. Sometimes questionnaires are printed in a newspaper or magazine with the suggestion that readers answer the questions and mail the form in.

The interview form of survey, either face to face or over the telephone, has the advantage of flexibility. The interviewer can adopt or change the questions according to the response of the person being interviewed. He can also ask for, and usually get, more kinds of information than can be requested on a questionnaire, for people will answer some questions orally that they will not answer on paper. Finally, a trained interviewer can often sense attitudes and opinions of the person being interviewed, even when these are not put directly into words.

However, that flexibility can be a disadvantage if the interviewer isn't thoroughly trained in his work. If an interviewer misinterprets any answers he gets or makes any mistakes in recording information, those mistakes will appear in the final data, and bias this data. There is more chance for human error in the interview than there is in the questionnaire method.

Telephone surveys are becoming increasingly popular. It is usually the least expensive of the interview techniques, and has become increasingly easier to use because of the services now offered by telephone companies. However, lengthy questioning usually cannot be handled over the telephone, and there are many kinds of information that people may be willing to discuss face to face but are unwilling to discuss with an anonymous telephone voice, especially those questions that deal with income and material wealth, for fear of robberies or burglaries.

Questionnaires have some important advantages. By using the mail, a broad range of people in many geographic areas can be surveyed at the same time. In addition, people who fill in questionnaires usually take the time to consider their answers carefully, and they seldom worry about wording their answers to impress anyone. However, questionnaires are an excellent technique provided that enough people reply to make their answers representative of the group as a whole. Many companies consider themselves lucky to get a 10 percent response. In addition to getting enough people to answer, there are the problems of compiling or finding good mailing lists, of setting up good questionnaires, and of paying the steadily increasing postage rates.

When to Survey. Choosing the survey target can either be a very simple job or a very complex one. The choice will make or break the validity of the data obtained. If everyone in a group is surveyed, as in the case of the members of the trade association or the charge account customers of the store, there is no problem. If a sample has to be picked, the task becomes more difficult.

A sample is a limited selection of persons representative of a larger group. Various techniques are used by researchers in choosing samples. One of the most common is random sampling, the method of selection in which each one of the total group has an equal chance of being included. Choosing names by chance from the local telephone directory is an example of random sampling. Nonprobability or weighted sampling is the method of selection that deliberately includes people with specified characteristics. A very simplified example might be that of a manufacturer with 50 wholesaler customers and 100 retailer customers; the manufacturer in this case might weight the sample to contain two retailers for each wholesaler surveyed.

Some companies prepare their own samples; others depend on professional list-compiling companies. Professionally compiled lists, of course, are used almost always for mailed questionnaires or mailed advertising, not for interviewing purposes. List companies can put together practically any kind of list a company can imagine. The price a marketer pays for the use of such a list (for lists are usually rented rather than sold) depends on how specialized the list is. For instance, a list of people who have bought houses within a particular area within the past two years would be relatively easy to compile, and therefore relatively inexpensive. But if that list were pruned to contain only the names of those with net incomes above a specified level, and who purchased that house as a second home or summer home, the list would cost a marketer more—and be more valuable, if that is the kind of data the marketer wants.

The Information. A survey may be seeking data either in the form of hard facts or of opinions or attitudes. Facts are much easier to collect, but opinions and attitudes can be useful, if they can be collected carefully and acted upon early enough. For instance, facts about what customers in a particular market are buying today are useful for determining immediate and current planning, but opinions about what customers will buy tomorrow, if those opinions are valid ones, can help the marketers determine their plans for tomorrow.

Obviously, the questions asked in an interview or on a questionnaire must be carefully prepared. They must be clearly stated and easily understood, yet not appear to pry or demand. Questions must be easy to answer, and interesting enough so that the person being surveyed will want to answer them. They must be in logical sequence, so that one question leads naturally into the next and one does not influence the answer to the next question. Most important of all, the questions must be phrased in such a manner as to obtain the kind of data the marketer is interested in without encouraging bias in the survey answers.

The flexibility of the interview allows the interviewer to explain or rephrase a question if the person being interviewed seems puzzled by it, as long as the interviewer doesn't bias the question in rephrasing it. It is also possible to learn much more about opinions and attitudes in an interview, since people will talk more freely and at greater length than they will when writing questions on paper. The face-to-face interview is the best technique for drawing opinions and ideas from the person being interviewed, and often such an interviewer is given considerable leeway in choosing the best way to present the questions. In contrast, the interviewer is more likely to be expected to adhere to a set list of questions in a telephone interview in order to complete the interview within a specified amount of time.

A questionnaire must be completely self-explanatory. It must give instructions, encourage participation, and be easy to understand and fill in. Although questions about opinions can be asked for on questionnaires, this form is most useful when there is a need to collect hard facts. The Federal Census form distributed every ten years throughout the United

States and wherever U.S. citizens are living is a good example of a hard-facts form. Age, income level, occupational category—questions like these leave little room for confusion.

Observation

What people say they do and what they actually do can be quite different. For this reason, and because it is the easiest and most accurate way to collect some types of data, the observation method is widely used for some types of research. The observation method of collecting data involves observing and recording actions that actually take place.

The main advantage of the observation method is that it records actions. What people do is recorded as it is being done, not as people remember doing it. What happens is recorded as it happens, not as it is remembered or judged to have happened. For instance, a person might estimate that he turns on his television set an average of three nights a week for a period of two to three hours each night. Observation might show that he actually turns his television set on an average of four nights a week, one night playing it for about four hours and the other nights playing it for an average of about an hour a night.

The observation method also eliminates weighting of the data, conscious or unconscious, that results from a prejudice or bias of an interviewer. For instance, an interviewer might find some questions to be asked dull or confusing and, thus, would tend to collect less data on these points during the survey, or an interviewer might take an instinctive liking to a person being interviewed and tend to give that person's answers more weight than they deserve. This is referred to an interviewer bias.

There are two disadvantages, however. The observation method can be relatively costly, as it is likely to tie up trained people and expensive equipment for long periods of time, much of which is spent waiting for something to happen rather than recording actions. In addition, opinions and attitudes, so important in marketing, cannot be recorded, since they are states of mind, not actions. Two basic forms of the observation method are frequently used in market research—the personal and the mechanical.

Personal Observation. Personal observation is particularly useful in various facets of retailing. A store often uses "shoppers," trained observers who shop different departments of the store and observe and record the behavior and actions of salespeople. A retailer considering opening a shop in a shopping center may copy down the license plate numbers of cars parking at the center during one week to determine the area from which the center is drawing its customers. A department manager may station a trained observer near an unusual display to observe and record the reactions of people who stop to examine the display. In each case, the technique records hard facts in the form of actions, resulting in a collection of data that probably could not be gathered in any other way.

Mechanical Observation. Mechanical observation involves recording machines. These machines "observe" and record data, in the form of impulses, photographs, or mechanical counts. The cord stretched across the highway feeds an impulse into a recording unit every time a car passes over the cord. It provides an excellent, unbiased record of the number of automobiles that use a specific route during a specific time period. The audimeter, used by various market research firms, records the exact time a television set is turned on and off, as well as what stations it is tuned to. In many stores, cameras and remote television sets are used to analyze the buying patterns of customers and monitor the flow of traffic in the various departments. It has the added advantage of using the equipment to guard stockroom areas and merchandise displays to prevent theft. Banks, too, now use cameras to record all action that takes place at each cashier's window, so that there will be a photographic record of any attempted hold-up. When a bank criminal is caught, he may be able to throw some doubt on the cashier's identification of him, but he can't explain away a filmed record of his crime.

Experimentation
The experiment is a technique that involves changing one or more factors to show the effect of change under controlled conditions. Marketers probably become involved in more experiments than they realize. Some of these are laboratory experiments, in which product production methods or product qualities are measured and tested. But that laboratory can also be the marketplace, where primary data can be collected by experimentation.

The set-up for an experiment requires an experimental group and a control group. A change is introduced in the experimental group, but not in the control group. After a period of time, both groups are examined to see whether the change has caused any variation between the experimental group and the control group and to determine what that variation is.

Application in Marketing. Aside from the use of experiments as a method of collecting technical data about a product, the major use of the experimental method in marketing is to test variations in product, price, promotion, or distribution under actual market conditions. Experimentation is an important step in the introduction of a new product—as well as in helping to determine what changes are needed in an established product to keep it profitable.

For instance, a company might decide to strengthen its marketing mix by changing the packaging of one line of products. This is what it could do: One, perhaps two, new types of packaging are developed and produced in a limited quantity. Three market areas, which are as similar as possible in all major respects, are chosen. The two new packaging designs are introduced in two of the areas, one in each area, and the third area is designated as the control area. After a set period of time, the company examines both the sales records produced by all three areas plus the customer comments that have been collected by salesmen. The resulting data might very well show that one new packaging idea produces considerably more customer interest than the other new design and more interest than the old design.

Sometimes an experiment has unusual results. The following is a simplified and edited version of what an experiment taught one marketer: When the marketer, who ran a men's shop with heavy emphasis on high-priced merchandise, decided to clear some excess inventory of summer shirts off his shelves, he marked some of them down 30 percent and some of them down 60 percent. His reasoning was that the few marked down 60 percent would be particularly attractive price leaders that would bring customers flocking to the store. When the sale was completed, he found he had sold most of those shirts that had been marked down 30 percent, yet he still had an excess of shirts marked down 60 percent. The data, when supplemented by a quick survey of the salespeople and a few customers, showed that the customers didn't trust too low a price, fearing that the product was inferior.

The Pros and Cons. The major advantage of experimentation lies in its realism. It is the only one of the three methods of collecting primary data that can be used to test a factor under actual or simulated market conditions. However, it often is a long time before sufficient data can be collected to make the results usable, and often the conditions of an experiment are very hard to control properly enough to make those results valid. Thus, while experimentation is a vital method for getting the reactions of the marketplace, it also can be costly and time-consuming, and may lack validity and reliability.

SEQUENCE OF ACTION

Collecting information involves a series of steps, each of which must occur in proper sequence. When this sequence is followed, then the marketer has the most likelihood of making the right choices and ending up with the data that will help him to make the right planning decisions. First, there must be a recognition of the need for data. Next, there must be a definition of the kind of data needed. Finally, there must be a choice of the methods to be used in collecting the data needed.

Recognizing the Need

When a marketer accepts the need for good planning, he or she must also recognize the need for data on which to base planning decisions. Some data is generated as part of normal business operations and thus forms part of company records. All the necessary data must be brought together, and the collection process can be set in motion only if the marketer recognizes a specific need.

First, a marketer has a regular and continual need for general operating data. This represents the steady flow of information that must keep coming across a marketer's desk if he or she is to keep in touch with all the facets of the business and do the job properly.

Second, a marketer frequently has a need for special data, the kind that will make for the right decisions when a specific problem or opportunity arises. For instance, general sales figures are sufficient as long as sales remain within the planned pattern. However, if there is a sudden change in the sales of a particular product, specific data is needed to help the marketer make the right decisions. Information might be needed about competing products, new trends in the market, and any changes in the product itself, to name just a few of the possibilities.

Defining Data Needs

Once a marketer recognizes why data is needed, whether for general business operating decisions or for help in planning the way to handle a specific problem or opportunity, the next step is to define the kinds of data needed. Sometimes this is a simple task, particularly if the planning involves a routine decision. More often it can be difficult. Many kinds of information might be relevant, but collecting all of it would be far too expensive and time-consuming. What must be done is to sort the essential kinds of data from the nonessential kinds of data. Here is the best method.

First, identify all the kinds of data that might be useful. If the data is needed for a routine planning job, this will not present a problem, for the same job of identification will have been done over and over again in the past. If the data is needed for a new or unusual problem or opportunity, then the identification of the possibly useful data will take more time.

Next, identify the data that will be most useful in making the needed decisions and eliminate all consideration of both obviously insignificant and probably insignificant data. Time is an important consideration in making these decisions, so a time limit has to be set for choosing which types of data will be useful and which types probably will not be useful. If there is still some indecision about some kinds of data at the end of that time limit, it is probably better to discard those uncertainties as being "probably insignificant." Cost also plays its part. Some kinds of data might be very useful, but far too costly to collect. These, too, should be eliminated.

Choosing the Methods
Once the kinds of data to be collected have been established, the final decision involves the methods to be used to collect that data. Sometimes the entire market research effort will concentrate on a single method of collecting information, but more often a mixture of methods is used. As always, the aim is to get the most useful information for the least cost in time and money.

The best way to choose among the various collection methods is to consider what and who, and then to compare the choices with the resources available. *What* in data involves the kind of information to be collected: hard facts (age, income level, or car ownership), opinions (attitudes, awareness, prejudice, bias both for and against), and behavior (actions, what actually takes place, what is done). The *what* types of information can be collected both as secondary data and as primary data. The most important points to remember about *what* data are that:

- Hard facts are usually verifiable and thus easy to work with.
- Opinions are impossible to verify and, no matter how carefully collected, are always open to question and debate.
- Behavior is completely verifiable, since it is collected only as primary data through the method of direct observation.

This does not mean that any one kind of information is better than another. For instance, opinions, although impossible to verify, can be very useful in plotting possible future trends in the market. On the other hand, yesterday's and today's sales figures also give an indication of what tomorrow's sales may be like, and sales figures are hard facts and thus usually verifiable. Therefore, if collected opinion suggests one course of action and collected sales figures suggest another, the verifiable data may have more meaning.

Who in data involves people, either in their roles as researchers or as those being interviewed or questioned. This is the area in which the greatest amount of error can occur. If the researchers are not properly trained or are sloppy in their job preformances, the result will be sloppy data. If the sample of people questioned is poorly chosen, or if the people themselves are confused or irritated or distracted for some reason, the result again will be inaccurate data.

This simply means that the better control the marketer has over the human element involved in the collection of the data, the more accurate the results are

Part Two   Process   79

going to be. One trained researcher will outproduce three untrained researchers. A questionnaire sent to one small but carefully selected group of customers probably will produce more useful information than a questionnaire broadcast to an unweeded general list of people. Once a marketer has measured the data needed in terms of the what and the who, he or she can consider the amount of money, manpower, and time available, and choose that method or combination of methods that will give the best data for the least investment. The choices available are almost unlimited. The only guide to remember is that the investment in the collection of data, regardless of whether that investment is in time, money, or manpower, should be expected to produce a logical return. That return on the investment may be tangible or intangible, short-range or long-range—but a manager should be able to explain exactly why the expense is worthwhile.

DATA COLLECTION SYSTEMS

The data collected by a marketing firm usually falls into two categories: data used for control and data used for planning. Data used for control is the accounting data that is generated regularly as transactions take place. Even though these figures are useful for planning, their main function is that of internal economic control. They are the results of previous planning, and they are of immediate importance to a marketer only when they suddenly indicate that something unplanned is happening.

Because most marketing executives need every second of time they can find for working out plans for the future, they see relatively little control data. Instead, they choose to use the exception reporting technique. *Exception reporting* is that method of control requiring the reporting of accounting figures only when those figures do not fall within the planned or set minimum and maximum. Thus, with exception reporting, an executive who does not receive any accounting figures can assume that all is going according to plan—because if it isn't, he will receive an immediate alert. The time that he would normally spend checking through the control figures can be spent on advance planning.

Planning data is used by the company both to meet current problems and to take advantage of current opportunities as well as to carve out opportunities for the future. As mentioned above, planning data includes portions of control data. For instance, a manager will need details about a product's sales pattern and its promotion budget and promotion plan when charting what effort should be put behind the product in the future. However, planning data also includes collections of other kinds of data, anything that will be useful in making a planning decision.

All marketers collect data, from the smallest shop to the largest company. All marketers collect both control data and planning data, although they may not differentiate between the two. The data-collection system of a small shop may be an accounting book and a sheaf of sales slip carbons, whereas the data-collection system of a large company may be a complex of people and equipment that requires an upkeep of hundreds of thousands of dollars a year. The efficiency and value of both depends on whether it gives management what it needs at a reasonable cost.

Data collection is necessary for good planning—and good planning is necessary for collecting data. It has to be planned, by small companies as well as large. A system should be set up to handle the needs of the company, to be reviewed and revised whenever necessary.

Small-Budget Data-Collection Systems
A system needn't be expensive to be effective. All a good business system requires is an established methods of doing something. Even the smallest marketer has systems for conducting business. The essentials of a good data-collection system are already part of even the small marketer's operations. The records that any marketer must keep for tax purposes form a solid core of useful data. These records consist mainly of control information: income, expenses, sales. By using this information as a base, it isn't hard for a small company to set up a good system of collecting some planning data as well.

First, a marketer should examine the data he already is collecting to see how much of it should be useful for planning as well as control purposes. The marketer then can arrange to have this data sorted regularly, so that whatever control data he needs to see regularly he sees, and whatever data would be useful for planning is collected and held until needed. Such a sorting might simply involve having the company's office clerk or accountant spend a couple of hours each month sorting and copying figures.

Next, the marketer should investigate his needs and see what other kinds of data could be collected regularly that would be helpful in his planning. For instance, if sales data is being collected on the basis of total sales per broad merchandise category, would it be valuable to break that category down into smaller merchandise classifications and find out exactly which products are producing what profits? The cost of collecting the data has to be weighed against the value of the data—but the more adeptly a marketer learns to use data, the more value he usually places on it. Thus the marketer who says, "It isn't worth it," ought to think again; is it really an unwise investment—or is it a matter of his not knowing how to use the results he will get?

Finally, the marketer should investigate methods of collecting other data than is generated by his own company's transactions. Would it be useful to subscribe to one of the market reporting services? How about making a regular check of the business statistics collected by local sources? Would a once-a-year questionnaire to regular customers produce useful data? How about hiring an interviewer to talk to customers and find out exactly who shops at the store?

The small marketer, with relatively limited data needs compared to some of the huge diversified marketing companies, will find that there are ways he can collect all the data he needs for a very reasonable investment. The important thing is for him to set up his system and adhere to it, and then use his data as wisely as he can. This is the best insurance he can have against the possibility of a competitor cutting into his business.

Large-Budget Data-Collection Systems
For many years, even the largest companies had data-collection systems that simply had been expanded as the company grew. A company usually starts out by collecting only routine accounting reports. Next, it

adds some sort of sales analysis reports to the list of data collected regularly. Finally, a company usually decides it needs to get data from outside sources on a regular basis. Thus, the data system is built up piecemeal, on an as-needed basis.

A system that has expanded with its needs usually produces the data the company wants, but the collection process tends to be more cumbersome and more expensive than it need be. The equipment that was adequate ten years ago may no longer be the most economical and efficient for today's needs. The sources that were chosen ten years ago may no longer be the most fruitful today.

During the 1960s, the importance of data collection and analysis became increasingly obvious to the marketing world and the term "marketing information system" became a part of the trade language. A marketing information system, or MIS, is a set of procedures and methods for the regular collection, analysis, and presentation of information for use in making decisions. A good MIS gives a marketer the information he needs, when he needs it, in the form he needs it. A good MIS produces regular reports on schedule and special reports when needed.

Equipment. In a larger marketing operation, the MIS may include a communications network that stretches across the country, a central complex of electronic equipment, and the full work time of 50, 60, or 70 people. Large companies have been able to rethink their entire data needs and to redesign their entire data systems primarily because of the kinds of equipment now available. The manager of a distant store used to be able to telephone his week's sales results to the home office and think that was speedy communication indeed, but today it is possible to record details of each transaction right at the cash register, send those details by leased line to the home office, and have them fed right into the central computer, all automatically.

Today's electronic equipment gives companies the ability to collect, store, and analyze data much more easily and quickly than ever before. Electronic equipment cannot be considered inexpensive, even for the largest of companies, but it is certainly less expensive than it once was, and many companies feel that its efficiency more than justifies the investment. The benefits of electronics are not limited to large companies. Even the smallest marketer can profit from them. There are companies that sell computer knowhow and lease computer time. Small marketers can buy just as much of each as they need and can afford.

The Human Factor. No matter how much money is invested in a marketing information system, in the form of equipment and workforce, a system is only as good as the thinking that went into it. This point probably is forgotten more often in large companies than in small companies. In a small marketing operation, the manager knows that his decisions determine the course of the company's operations. In a large company, a manager may tend to depend too heavily on "the system"—forgetting that it is his thinking that must both create and make use of the system. A management information system, regardless of size, is as good as the planning that went into it and the caliber of the executives who use its data.

The greatest problem and the greatest asset in the collection of data is the human factor. People, not equipment, must recognize the need for data, pick the kinds of data to be collected and the methods to be used, and do part of the actual collecting and recording of that data. People, not equipment, must make the marketing decisions on the basis of that data.

Equipment is a vital tool in the collection of data, particularly for the larger company. But equipment is just that, a tool, an aid for the marketer. A good system is vital in the collection of data, in a company of any size. But like equipment, a system is just a tool, a way of helping the marketer organize his actions. The key factor in collecting data, as in all facets of planning, is the marketer himself: what he knows, what he learns, how well he uses the tools available to him, and what decisions he makes.

Chapter 5
Tools for Analyzing Marketing Data

The bits of rock brought back from the moon's surface don't look particularly unusual or exciting. It is their source that causes a crowd to gather and stare when any one of the rocks is put on public display. Nor would the actual raw results of one of the tests being performed on those rocks be any more fascinating to the ordinary person, for raw results are simply a mass of figures and symbols.

Yet those rocks are a rich source of data, and the tests being performed on them are putting that data into usable form. Only when those test results are properly organized and analyzed does the fascinating and valuable information they contain become clear and understandable.

Any collection of raw data is a jumble. Many collections of data made by marketers are even more complicated and nebulous than the figures produced by the moon specimen tests, for marketing data includes not only facts, which can be verified, but also opinions, which can never be checked. Regardless of what is contained in a collection of data, however, only when its contents are organized and analyzed will it yield the useful information that a marketer needs for his job of planning.

As discussed earlier, market research consists of two activities: the collection and analysis of data. The previous chapter explored means of collecting data. This chapter covers the analysis of data and how the results are used in planning.

The first task in analyzing a collection of raw data is to prepare that information in a form that can be analyzed easily. Accuracy is checked; the facts and figures are classified in meaningful ways; statistical manipulations are made so that the relationships between groups of data are as clear as possible.

After the data is prepared, the most appropriate tools of analysis are applied. These tools may be in the form of performance measurements in various areas of operation. They may be financial checks, such as return on investment or the break-even point. The results of the various analyses are then used as guides in forecasting and budgeting, which are basic operational prerequisites for the preparation of any individual marketing plan.

Through the analysis stage, the work involved is part of the formal discipline of market research. It involves the use of statistical techniques, which are exacting in the way they must be handled. Forecasting and budgeting require other managerial talents as well: logic, common sense, a good business background, and the ability to project today's facts into the action to be taken tomorrow.

PREPARATION OF DATA

One retailer had sent out a detailed questionnaire to her 500 charge account customers. Since her operation was only of moderate size, and since she wanted to keep close track of what was happening, the questionnaires were to be returned directly to her office. Her secretary was to collect them and, at a specified cut-off date, send them to the statistician hired to organize and analyze the data.

The questionnaires began to come in, and every afternoon, the retailer asked to see the questionnaires that had arrived that day. Over a period of two weeks, to the point of the cut-off date, a total of some 100 questionnaires were returned. The retailer read each reply, spending an average of five minutes on each one. She enjoyed reading through the answers—and that enjoyment cost her 500 minutes or more than eight hours of her working time. All she got in return for that time was an idea of how some half-dozen particularly articulate customers had answered, a very few people whose answers may or may not have been typical of her customers as a whole.

Moral: Examining raw data can be fun and almost irresistible, but it should be considered a time-consuming luxury that does not take the place of proper analysis. Occasionally it can be misleading as well as time-consuming, for an unusual answer may create a bias in a marketer's mind, even though a formal analysis of the data shows the answer to be one of a small minority opinion.

Proper preparation of data is essential for good analysis. Trying to analyze unprepared data is like glancing at pieces of a jigsaw puzzle at random: it can be amusing, but it doesn't put the puzzle together. Preparation involves three steps: a quick appraisal of the accuracy of the data, an arrangement of the data into proper classifications, and a manipulation of the data to make comparisons easier.

Checking for Accuracy

With all the emphasis placed on the care needed in collecting data, it would be easy for a marketer to take it for granted, once the collection process is complete, that it has been done properly and that all the data contained in the collection fall within predetermined limits of accuracy. However, every marketing process should include more than one built-in check as a safety control. This is the final stage at which the accuracy of the raw data can be checked. Once the facts and figures are grouped and merged together the individuality of the separate bits of information disappears.

A check for accuracy at this point usually involves two areas. First, the way the collection was actually made is reviewed to make sure that what was done was in line with what had been specified by the marketer. Second, a few of the verifiable facts or figures themselves are checked back to their sources for accuracy. If the proper collection methods were used, and if the few random figures selected for source-checking turn out to have been faithfully recorded, the marketer can assume that the collection probably contains accurate data.

Let's say a particular collection of data was used to make a survey of wholesaler customers and to check the latest Department of Commerce figures in a specific area. To test the accuracy of the data, the marketer or researcher might check back to see that questionnaires were sent to all the wholesaler customers, that a reasonable number of the questionnaires were returned with enough of the questions answered to produce a valid sampling of the group as a whole, that a half-dozen of the most important figures recorded from the Department of Commerce were the most recent available, and that they were recorded without introducing error.

This is also the last chance to review whether the raw data is reliable and representative. Reliable data can produce trustworthy and useful information, whereas mistakes in collecting and recording data will be perpetuated all the way down the line, right through the analysis of the data. The result will be distorted information that can mislead the marketer and weaken the effectiveness of his or her planning.

Classifying the Data

After the accuracy of the raw data has been given a final check, the data must then be arranged according to whatever classifications permit the best analysis of the collection. Exactly what those classifications are will depend on the purpose of the analysis and how the resulting information is going to be used in planning.

For instance, the mass of collected data about the transactions that occur in a particular store department in a given week could be classified by dollar amount of transactions, by number of items involved in transactions, by category of each item purchased, by the salesperson involved in transactions, by time the transaction took place, and by various characteristics of the customers making the purchase, among many other possibilities. A manager setting up a work schedule for the department would be very interested in classifying the transactions according to the time they took place in order to staff the department to meet probable peaks in customer traffic. A manager interested in figuring out profitable future merchandise assortments probably would concentrate on the classification of transactions according to the category of item purchased.

Types of Classification. Although it is possible to classify data in a great many different ways, each classification falls within one of four basic types of classification:

- *Quantitative*: a classification of data by difference in amount, such as a classification of sales by number of units sold or number of transactions involved;
- *Qualitative:* a classification of data by difference in kind, such as a classification of sales by type of product sold, type of brand sold, type of customer involved in sale;
- *Chronological*: a classification of data according to a time sequence, such as a classification of the average number of transactions handled during each hour of the working day;
- *Geographic*: a classification of data by difference in location, such as a classification of sales by store location or market area or sales district.

The science of pure mathematics recognizes only two types of classification of data: quantitative and qualitative. Chronological and geographic differences are considered either quantitative or qualitative. But to marketers, for whom timing and location are so important, chronological and geographic classifications are often ranked as separate and equal in importance to the two basic types recognized by the mathematical world.

Identification of Classifications. A classification, when it is put down on paper, usually takes the form of a table, chart, or graph. It is very important that these be properly identified. The identification should include a definition of what the table, chart, or graph is about, an explanation of any unusual factors that influenced the findings, and an identification of the amount of data on which the table, chart, or graph is based.

For instance, a table of shoe sales classified by size of shoe could be very useful in helping the store manager determine his future inventory needs. But what if the table includes figures both for "sports shoes" and "casual shoes"? In order to analyze and understand the information, the manager has to know exactly which types of shoes are included in each category. If he doesn't know, he may end up stocking his store with sneakers, when what he really needs are more loafers. What if the data was amassed during the back-to-school period, a traditionally strong period for shoe sales? Unless the manager knows what type of shoe was sold and that the collection of data was made during that particular time period he may find himself overstocked with school shoes, but understocked for other kinds of styles during the winter and spring seasons.

What happens when the table represents only a small sampling of sales, such as a week's actual figures projected for an entire month, or when the sales were made to a small random sampling of customers? Small samplings sometimes point up possible situations that a marketer might investigate, but only samplings large enough to represent the group as a whole can be considered reliable trend indicators.

Utilization
The final task of data preparation is the utilization of the data to make relationships as meaningful as possible. This utilization involves the application of specific statistical techniques to the data so that analysis can be done more easily. The most common

kinds of utilization involve working out averages and medians, percentages and ratios.

A ratio is the relationship between two figures reduced to the form of a divider and a dividend. Stockturn, which is the number of times an average inventory of stock is sold during a given period of time, is usually expressed as a ratio. Products that move rapidly, such as canned food, may have stockturn ratios of 10:1 or 12:1 per month, which means that the basic stock carried by the market sells, is replaced, and is resold 10 or 12 times during a given month. In contrast, fine jewelry may have stockturn ratios more in the neighborhood of 2:1 or lower in a six-month period, which indicates that the items involved earned only two sales or fewer during that period.

In addition to underlining relationships between figures and to making figures easier to analyze, utilization often results in a useful condensation of the data contained in what otherwise would be a very long table. Instead of having to work with 52 weekly totals in a table that shows sales for a year, it is possible that a single weekly average or median figure for each selling season or distinctive sales period could be used for much of the analysis. As long as the summary or condensed figures are representative of the original groups of figures, they can simplify and speed up the analysis.

PERFORMANCE ANALYSIS

The overall analysis of a collection of data almost always involves a series of separate individual analyses. Each one of these analyses is designed to explore one particular facet of the data. Thus a collection of sales figures might be analyzed in terms of types of items sold as well as the share of the market won, in terms of gross margin as well as selling costs.

Remember, a manager usually needs two kinds of data as an aid in making decisions. He needs general operating data to help him with his normal business planning, and he needs special data when he has to make decisions concerning specific problems or opportunities. Because the collection of general operating data is a regular and continuing process, the analyses of this data are most frequently performed by the marketer. The most basic of the analyses performed on general data involve calculations made of such performance records as sales, market share, distribution, profit and cost, and the selling force.

How frequently such analyses are made depends completely on the kind of business that is involved. A manager of a small business or one handling a very stable product may not need to check performance records more than monthly or quarterly, and may not need to make any particularly detailed analyses more often than quarterly or seasonally. A manager with a variety of products or a manager handling highly volatile products needs to keep much closer track of performance records. Sometimes the entire market picture can change unexpectedly within a week or two, and that change can be significant enough to require immediate attention and action of the manager.

Sales Analysis

A good sales analysis does not show only the developments in total sales volume. It also helps the manager pinpoint weaknesses and strengths in the sales picture as they occur. This kind of analysis should do the following: (1) classify sales by product or category, (2) add up unit and dollar totals, and (3) compare totals with previous figures or with planned figures to determine whether results are behind, even with, or ahead of expectations.

The key to success is making sure that the sales figures are divided according to the classifications or categories that the manager will find most useful. Depending upon the kind of business involved, a manager may need two classifications or 100 classifications, they range from geographic region to type of fabric—but only when the figures are divided into meaningful groups will the results of the sales analysis be truly useful both as a planning tool and a control tool.

Many larger industries have had to rethink their entire classification system. In any strongly competitive market, it is important that a manager plan on the basis of customer wants and needs—how customers buy—not in the more traditional terms of how the marketer obtains or displays the merchandise. It is for this reason that the retail industry has worked out a new classification system that cuts across many of the older departmental and category lines.

The need for detailed data has also been a problem for the larger company, simply because of the amount of data that has to be collected for a good sales analysis. However, the new automated equipment has helped many a manager collect the data that today's markets require. Both mechanical and electronic equipment have automated and speeded up the collection, preparation, and analysis of sales data, making it possible for larger amounts of data to be handled in a very small fraction of the time that manual handling of smaller amounts used to require.

Sales analysis is not quite as easy as it would seem to be—at least, a good sales analysis isn't, but sales are the heart of any business operation and money and time invested in a good sales analysis are money and time well spent.

Market Share Analysis
Those marketing companies large enough to be figuring sales based on their share of the total market usually need some sort of market-share analysis. In a way, a market-share analysis is a test of the sales figures. A 10 percent increase in sales may look very healthy, but how healthy is it if the size of the total market has increased 20 percent during the same period? That would mean that the company's share of the market has not increased or even held even, but has actually decreased, so far as the company's competitive position in relationship to other companies is concerned. A share of the market analysis is figured this way: first determine total market sales by dollars and units, then determine company sales by dollars and units, and finally, establish the percentage of company sales in relation to total market sales.

Thus, while company sales figures are available from within the company, a share of the market analysis also requires figures collected from outside the company. But if a marketer is interested in a market share analysis the necessary figures are collected from external sources just as regularly as sales figures are collected from internal sources.

If a manager decides to undertake his own collection of market figures, the job can involve expensive and lengthy research in many fields. However, in those marketing fields in which companies are most likely to be interested in making a market share analysis, market figures are often available from various syndicated data firms. The cost of getting the data on a regular basis may not be cheap. In some fields, a steady flow of data about changing market statistics may cost a company upwards of several thousand dollars a month. For those companies that plan in terms of market share such information is well worth the cost.

The companies most likely to be interested in the share of the market analysis are those with a measurable share of one of the big consumer markets with a high stockturn rate. In such markets there frequently is a shuffling of product preference and customer demand, and a marketer must continually adjust his operations to meet these changes or else lose valuable business.

Distribution Analysis
While the sales analysis and the share of the market analysis can show a marketer what is happening, the distribution analysis can explain why it is happening. Manufacturers and wholesalers are particularly interested in the distribution analysis. Single-outlet retailers seldom are interested, but multi-unit retailers often include some variation of a distribution analysis as a part of their regular checks.

A distribution analysis concentrates on the outlets selling the company's products. There is no set formula or method of procedure. Each company must determine what information it needs and then collect and analyze that information regularly. Typical information might include:

- Type and size of outlet, location of outlet.
- Sales of company's products generated by outlet, in units, dollars, net profits, and in total sales made by outlet (if possible).
- Inventory carried by outlet.

- Shipments made to outlet, receipts acknowledged by outlet, average time stock remains in outlet.

With such information, a marketer not only determines which outlets are doing the best job with the company's products, but also spots possible weaknesses or trouble spots, such as delays in getting shipments to an outlet, for instance, or an unusually heavy inventory beginning to build up in a particular outlet. A marketer uses the distribution analysis both to control current operations and to plan future operations. When a trouble spot appears, it can be investigated and corrected. When a single outlet or a particular kind of outlet shows unusual success, the reason can be identified, and what is building sales can be encouraged in other outlets, if it seems logical and practical.

Most of the data needed is in the customer records and the sales and audit reports supplied by the outlets. Whatever else is needed can be collected on a regular basis from the outlets.

Profit Cost Analysis
Even if the sales figures look healthy, the share of the market figures look satisfactory, and the distribution analysis doesn't show any problems, the company's books can still show a net loss. Somehow, there is simply too much money going out and not enough coming in. This can be true for the company as a whole, or for one particular division within the company, or for one product or group of products. An analysis of profit and cost tells the story.

There are several ways that profit and cost figures are collected and analyzed. These include the gross margin system, the natural expense system, and the functional cost system.

Gross Margin. As a reminder, a gross margin system of profit and cost accounting compares all cost allocated to a department, category, or product classification with the gross profit indicated by the margin percentage. Here is the calculation:

$$\text{Gross margin \%} - \text{total costs \%} = \text{profit \%}$$

The gross margin is the final selling price less cost of goods sold. But a simple departmental analysis can leave some problems and opportunities undiscovered. Marketers with a variety of products and a number of selling departments should try to develop profit and cost analyses for the major product classifications. This can be hard to do at times, both because it requires collecting considerable data and because some general departmental costs are hard to divide among the various products carried in the department. However, only a detailed analysis by major product lines or classifications can show the marketer which products are the strong profit-producers and which products are not profit producers.

Natural Expense. The natural expense system is the standard method of business accounting. Costs are divided into three broad groups: materials, labor, and overhead, and then into various subgroups. Profit is the percentage of income earned in excess of these costs. Whereas the general method of figuring is valid for all kinds of business, the three standard classifications of costs are not always convenient for all kinds of marketers.

For instance, a large association may assign office floor space, and then charge divisions for that space in accordance with the number of people in each division and the jobs they do. Contrast that with the department store, where the amount of sales per square foot of selling space earned by each department is an important factor in determining how much space to assign to a department and what to charge that department for the space.

Functional Cost. The functional cost system attempts to assign all costs incurred by the company to a particular product or project. Those costs are then compared with the income produced by that product or project, and the result is the amount of profit produced. The basic formula is similar to that of the gross margin system:

$$\text{Income} - \text{costs} = \text{profit}$$

Or, in terms of planning:

$$\text{Planned profit} = \text{planned income} - \text{planned costs}$$

When any one of those planned figures is not matched by actual results, the marketer has to do some quick adjusting to keep operations in balance. The danger of a drop-off in planned profit or planned income is obvious, but even a drop-off in planned costs can be a warning signal. Unexpectedly lowered expenses could mean that something that should be done is not being done and that sales figures might eventually be affected.

A functional cost analysis is perhaps the most useful kind of profit and cost calculation a marketer can have, but it can be hard to figure. For instance, it is easy to calculate or record the cost of manufacturing and shipping a product, but it is far more complicated to estimate the costs of product development and improvement and then allocate them accurately among the different stages of the product's expected life cycle.

Part Two   Process

Sales Force Analysis

Among marketers, an important way of checking the efficiency of the sales force is by analyzing its selling cost as a percentage of sales volume for a particular period. Typical periods are a month, a season, or a year. Here is the formula:

$$\$ \text{ sales volume} \div \$ \text{ sales force expense} = \% \text{ sales force expense}$$

The sales volume may be the total sales income of the company as a whole, or of a single department or sales district within the company, or of a product or category of products. In the same way, the sales force cost may be the costs incurred by the company's entire selling staff, or by the salespeople within one department of a sales district, or by each individual salesman.

Retailers usually analyze selling force expense by total company sales in a small operation and by department in larger operations. Manufacturers, wholesalers, and industrial marketers are likely to analyze selling expense by sales district, by individual salesman, and sometimes by product or product category.

When individual sales performance is vital to a marketer, as it is in any operation that depends heavily on the sales force to do the largest part of the selling job, then such an analysis usually includes, in addition to the basic data in the above formula, such information as unit sales made by each salesman, expenses incurred by each salesman, gross margin earned by each salesman, number of calls made per day, number of new accounts obtained, and amount of time spent in nonselling duties. This enables the marketer to pinpoint which salesmen are contributing the most net profit to the company, which salesmen are finding the most new contacts for the company, and which work patterns seem to be most effective.

Thus, such an analysis is one of preparing and checking percentages and ratios, with salespeople and sales districts or selling departments being compared with each other, with planned goals, and with the figures for the company as a whole. In this way, both weaknesses and strengths in the selling force can be identified, the strong rewarded and the weak, if possible, trained and encouraged to improve their performance.

FINDING THE FINANCIAL STRUCTURE

The various types of performance analysis examine what has been accomplished and give results that serve both as a means of checking progress and as data for planning and decision making for the future. Some of the results are used in two important types of financial figuring—the analysis of return on investment (ROI) and the break-even point analysis. Both are used to project estimated figures into as accurate a picture as possible of what those figures will mean in terms of a future balance sheet.

ROI can be used to study the company as a whole or to study a particular project or product. It involves plotting the total cost investment that will be involved throughout the life of a product or project, and then working out what that investment will probably earn for the company in the way of a profit return.

The break-even analysis is primarily a test of future product profitability. The product may already be on the market, or it may still be in its earliest planning stages. The analysis explores the relationship between cost, price, and sales volume, showing where the break-even point in profit would logically occur in each possible combination of these factors. It can be used by the marketer to pick a specific mixture of cost and price and have a reasonable idea of what sales volume to expect and at what point the product will begin to see profits.

Return on Investment
Every dollar a marketer spends marketing must eventually pay off in a healthy return on that investment. It isn't enough that an investment merely produces a profit. That profit must be great enough, in terms of a percentage of the original investment, so that the company not only survives, but is financially healthy enough to compete and prosper.

The return on investment of money deposited in a savings bank is easy to figure, for it is a specified, guaranteed percentage per year. However, buy common stock with the same amount of money, and the return on that investment is harder to predict, since both the value of the stock and the amount of dividends paid will change as market conditions change during the year. Figuring the return on an investment by a marketer is even harder, particularly if the marketer is involved in a highly competitive and volatile field, and yet it has become an increasingly important calculation in making business decisions today.

An ROI calculation does not guarantee success. Conditions in any market can change too rapidly and too radically. However, a good ROI analysis does reduce the risks and guesswork of decision making. It can be used both to gauge the rate of return an investment is earning and to estimate what return would be earned during a future period by a specific investment. Thus it is the one form of analysis that takes into consideration the effect of time on the cost of money.

Gauging an Investment. Whether a marketer wants to analyze a past investment, a current investment, or an investment being considered for the future, the basic steps in figuring its rate of return are approximated the same:

1. List all the assets involved.
2. Figure the profit earned by those assets during the relevant period of time.
3. Figure the investment turnover for the same period of time.
4. Multiply the percentage of profit earned by the rate of turnover, and the result is the rate of return on the investment for the specified period.

For instance, a company may have two sales districts with identical sales figures, identical expense figures, and therefore identical profit figures. Yet the districts are quite different in terms of inventory carried and receivables outstanding. Therefore, the company may decide to do an ROI analysis on each sales district to determine which one is doing a better job with the company's money, as shown in Table 2. Although sales and expenses may be the same, producing the same profits on paper, district A is using a larger investment to produce those results than district B is. This means that district B is earning a better return on investment than district A.

Discounting the Cash Flow. As every marketer knows, money costs money. Borrow $1,000, and there is interest owed on that money each day until $1,000 is paid back. In accounting terms, interest, or the cost of

Table 2. ROI comparison of two sales districts.

CALCULATION	DISTRICT A	DISTRICT B
1. Sales	$200,000	$200,000
2. Cost of goods sold	−110,000	−110,000
3. Gross margin	90,000	90,000
4. Marketing costs	−30,000	−30,000
5. Net income	$ 60,000	$ 60,000
6. Accounts receivable	24,000	50,000
7. Inventories	66,000	20,000
8. Total investment	$ 90,000	$ 70,000
9. Profit on sales (profit margin) (line 5 ÷ line 1)	33%	30%
10. Investment (line 1 ÷ line 8)	2.2 times	2.9 times
11. Return on investment (line 9 × line 10)	66%	87%

money, "discounts" the full amount paid back to the amount borrowed.

An investment is a form of loan made by the company to the product or project. When a marketer is figuring the return an investment may produce in a future period, such as the next five years, he usually uses a standard compound interest table to figure the actual cost of the money involved in that investment. Thus the total assets to be invested each year are multiplied by the cost of the money or interest rate to determine total financial investment; then total expenses are deducted; and the result is anticipated net profit or net cash flow generated by the investment.

Thus, by estimating future returns on an investment in a way that takes the effect of time into consideration, a marketer can figure what rate of interest will discount the anticipated net cash flow for each future year back to the original compounded cost of the investment. This is an accounting task, and involves the use of a "value table" that charts amounts and rate of discount for each year into the future.

Break-even Analysis

A break-even analysis helps marketers answer questions like these: "What profits will a specific product earn at various levels of sales volume?" "If the fixed costs of producing product A go up, what increase in sales volume will be needed to maintain the same profit level?" "What effect will a 10 percent decrease in sales have on the profit level?"

The relationships between cost, profit, and sales volume are what a break-even analysis explores. If a product is sold for a specific price, then there is a definite series of relationships existing between the number of units that probably will be sold, the amount it will cost to produce them, and the profit they will earn.

These relationships are usually calculated by mathematical formulas and are then pictured in chart or graph form. The mathematics produces the results of the analysis, while the chart or graph helps dramatize those results.

Analysis: The *break-even point* is the amount of income or sales volume needed to produce profits equal to costs. Lower income or sales volume will produce a loss, for costs will be higher than profits. Higher income or sales volume will yield a net profit, for profit will more than cover costs, as shown in Figure 2. The formula for figuring the break-even point is this:

$$\text{Break-even point} = \frac{\text{fixed costs}}{\text{ratio of gross margin to sales}}$$

Figure 2. Break-even chart.

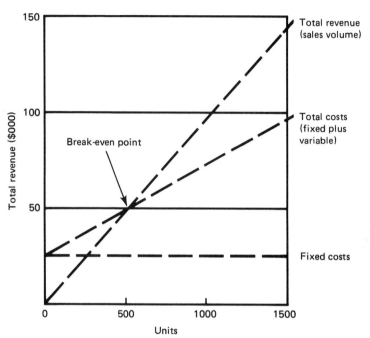

We can now examine some of the points on the total revenue curve and write them in tabular form. Let's say a marketer is considering adding a new line that is expected to produce the following figures:

	Dollars	Percent of net sales
Net sales	$100,000	100
Variable costs	−40,000	−40
Gross margin	60,000	60
Fixed costs	−30,000	−30
Net operating profit	$ 30,000	30

Fixed costs are estimated at $30,000, and the ratio of

gross margin to net sales is 60 percent. The formula then reads:

$$\frac{\$30,000}{60\%} = \text{break-even point, or } \$50,000$$

The Variables. The formula and the tabulation are worked out on the basis of a set price or series of prices being charged for the product line under consideration. Different prices will produce different results. In exactly the same way, a change in any one of the elements, in price, or sales volume, or fixed costs, will change the parts of the equation and the break-even point.

The marketer can experiment with variations of these elements to find the combination that will produce the results the company wants. Then, with the charted summary of the relationships before him, he can keep close track of how any changes in those elements will affect the overall profit and cost picture.

SOME PITFALLS IN ANALYSIS

Whether the analysis of marketing data is done by a trained researcher in a large organization, or whether it is handled by a small shop owner who does everything from opening the store in the morning to posting entries in the ledger every evening, there are some pitfalls that have to be avoided. The trained researcher may be more aware of these dangers than the small marketer, but both can be caught by them. What are the don'ts of analysis?

1. *Don't set out to produce startling results.* There is a temptation to try to read big news in what are relatively commonplace results. There is also a

temptation to bias the results of an analysis slightly to make the results seem "different" or "interesting." Something "unusual" is not what is needed by a marketer for his planning; instead, what he needs are solid, reliable deductions.

2. *Don't be swayed by favorable evidence, and don't ignore adverse evidence.* Often a marketer hopes that analyses will produce particular results. He wants the figures to show that sales are healthy, or that a new product is producing a reasonable profit, or that certain new equipment would be a good investment. In an attempt to have the figures show what the marketer wants them to show, an analysis may be handled, either intentionally or unintentionally, so that favorable evidence gets prominent treatment and adverse data somehow is lost in the shuffle. This may make the results look more cheerful to the marketer, but those results are just the product of wishful thinking, not a groundwork on which sensible decisions can be built.

3. *Don't get fascinated by details.* The value of an analysis is that it interprets and simplifies the original data. The whole purpose of an analysis is to produce summary information out of raw bulk details. Therefore, keep the results of each analysis as simple as possible, and do not let any analysis develop into a mass of statistics.

4. *Don't project samples that are too small.* Many people enjoy playing with figures, and these people sometimes have a tendency to draw remarkable inferences out of a very few pieces of data. While significant information should be recognized and noted, even when in a small minority, they should not have their importance magnified out of proper proportion. In addition, if insufficient data is available about

a particular subject, then any analysis of that data has to be tagged with the warning that the results are based on what would ordinarily be considered an insufficient sampling.

5. *Don't mistake opinion for fact.* Some data is verifiable; some isn't. Marketers work with both types. In some cases, however, the data used must be fact, not opinion, and substituting opinion for fact can lead to trouble later. In gauging the size of a potential market, there are a number of types of factual data that can be put together to make an estimate. This estimate is much more likely to be accurate than a statement in the press by a competitor.

6. *Don't put too much faith in averages.* While averages can be very useful, they are only a handy tool, not an accurate summation of figures. If the customers of a gasoline station drive an average of one-third of a sports car per family, it does not mean that every family will have an equal interest in sports cars. What it does mean is that there are enough sports cars owned by customers for the station manager to consider stocking a restricted line of special accessories.

7. *Don't confuse cause with effect.* Did the drop in dollar sales volume cause the increase in markdowns taken, or were the markdowns taken too early, causing the drop in dollar sales volume? Did sales of a particular line improve because a new distribution channel was chosen, or was a new distribution channel chosen because sales had improved? Misreading cause and effect can damage the validity of an analysis, which can be a real problem. In other words, when interpreting data, question, probe, and investigate until the results of the analyses make sense.

Part Two   Process   111

FORECASTING

The results of data analysis are the tools used by marketing managers to work out ways of handling their operating problems and opportunities. First, the marketer uses the data to forecast the probable future of the product or project under varying conditions. The job of forecasting is not a mechanical process, nor is it subject to the same kind of precise rules of mathematics that the analysis of data is. Instead, it requires a modicum of:

- *Objectivity:* The ability to keep one's own biases and prejudices separate from what the data shows.
- *Realism:* The ability to see both the problems and opportunities clearly, and to pick out the various types of information that will be most useful in making the right decisions about those problems and opportunities.
- *Competent use of formal logic:* For logic, as a base for all forecasting, keeps the marketer from making mistakes in interpreting what he knows and sees.
- *Business sense:* The ability to make decisions based on educated guesswork—the kind of guesswork that is backed by all the information available, common sense, and a firm grasp of the business of marketing.

Forecasting is a vital part of planning. Without a forecast of what can happen, no marketer can work out plans for the future. Without a plan, a marketer can merely react, not act. He always will be a little behind the competition, always in some danger of being edged out of business by a more farsighted businessman.

Timing. It takes the results of data analysis to produce good forecasts, but sometimes it also takes a certain amount of forecasting to produce the data needed for an analysis. Any kind of analysis that projects figures or performance into the future, as the return-on-investment analysis and the break-even analysis, can be used, based in part on forecast or projected figures. In turn, the results of an ROI or break-even analysis can then be useful in overall forecasting for the company or the market.

For example, a performance analysis may show that sales in a particular sales district are rising. A marketer may take the results of this analysis, add more data to those results, and work out a projected sales forecast for that district for the next three years. The results of that forecast may then be used in an ROI analysis to determine whether the increasing sales are going to mean an increasing return on investment in that district. The results of the ROI analysis may show that the rate of return probably is going to weaken, because the increased demand for the product will cause an unusually sharp rise in the expense of supplying that demand. Then the marketer might work out the final forecasts before planning and designating that particular district as a possible problem area rather than an area of opportunity.

Methods. Marketers have various means at their disposal for collecting data for forecasting, many of which are covered in the chapter on data collection. In addition to the more formal methods, a marketer can use shortcuts when trying to get enough information on which to base preliminary forecasts.

The manager who asks each responsible executive in the company, "What do you think the sales volume of the new tape checker will be next year?" is using a "jury of executive opinion" approach. It is one of the oldest of the informal ways of getting information, and is very similar to the majority vote method. The manager who questions salespeople, who are actually closest to the needs and opinions of the customers, or who questions the customers themselves, is using the "grass roots" approach. After all, the customers are the ones who make the final decisions of whether to buy a product or not.

Somewhat more along the lines of formal data collection is the "time series" approach. This method studies past performance and attempts to relate types of performance to specific business or economic conditions. For instance, is there a persistent relationship between the growth in disposable income and the sale of sporting goods, between minor depressions of the economy and the sale of cosmetics? If so, then spotting a change in the economy can herald a change in sales volume.

SETTING UP A BUDGET

Once a marketer has a good idea of what is ahead, then he must figure out, in general terms, what he will spend on meeting the challenges of the future. This is the task of budgeting.

A company usually prepares an overall budget, for company operations as a whole, and then prepares individual budgets for each section of the company that requires them. This chapter discusses the framework of a budget, while Chapter 6 describes how the parts of an individual budget are woven into a marketing plan. If finance or accounting is your strong suit, please move to the next chapter.

The major importance of a budget is its effect on planning and control. As a framework for planning, a budget specifies the income and expenses planned for the company as a whole and for each sector of the company. As a control tool, the budget is a series of benchmarks, enabling actual performance figures to be compared with planned figures and thus highlighting any variance from the plan.

A budget always must balance anticipated expenses against anticipated income, even though expenses can be divided into different categories according to their predictability. There are different types of budgets, but all budgets, regardless of type, must have certain characteristics if they are to be successful.

Types of Expenses

Some expenses are called "fixed," others are considered "variable" or "semivariable." The exact classification of an expense usually depends upon the kind of operation a marketer runs—but every marketer will have both kinds of expenses.

Fixed costs remain constant throughout the term or duration of the budget, regardless of changes in sales volume. These usually include executive salaries, rent, light, heat, and the like. Not only is there generally very little change in the amounts of these costs over the period of a budget, but also there is seldom any way to change these costs on a short-term basis.

Variable costs are those that fluctuate when sales volume fluctuates. They may include such items as the cost of order handling, of selling commissions, of selling expense, of advertising, of transportation and warehousing, among others. If sales drop, these costs may drop. A manager may find it wise to increase these costs in order to build up sales volume again. For instance, selling commissions automatically will drop if sales drop—but offering bonus commissions may encourage more effort from salespeople and result in a sales volume increase.

Semivariable costs are those related to production and sales, but not tied directly to sales volume. They usually include certain clerical and managerial salaries, and sometimes include transportation and warehousing, if these latter are not considered variable.

Any marketer preparing a budget must leave more leeway for change in his planned variable expenses than in his planned fixed expenses. Variable costs do change during the year, either as a direct or indirect result of changes in sales. A manager needs enough of a "cushion" in his variable estimates to give him room for maneuvering when necessary.

Types of Budgets

All companies have a master, or overall, budget. For many small marketers, this is the only budget needed. However, larger companies establish budgets for every phase of the company's operations that management feels needs financial guidance and control.

Thus the stationery store on the corner may operate solely by a master budget, and probably an abbreviated one at that, but the department store in the middle of the block may have four secondary budgets, one each for merchandising, publicity, store operation, and control, plus a number of departmental budgets feeding into each secondary budget.

A manufacturer, if relatively small, may supplement his master budget, with only an administrative budget, a manufacturing budget, an advertising budget, and a selling budget. A larger manufacturer may include a sales expense budget under the sales budget; budgets for production, materials, pruchases, labor costs, and manufacturing expense under the general manufacturing budget; and advertising, publicity, and general sales promotion under the promotion budget.

Budget Characteristics
Regardless of the kinds of budgets a company uses, every budget must contain several characteristics if it is to do a successful job of planning and control. First, a budget must be self-contained enough to cover only the area of operation for which the individual manager is responsible. It should not cover provisions for any income or expenditure over which the manager does not have direct authority. Second, if a company has a number of individual budgets, they all must dovetail neatly into the master budget. The master budget is the top management guide and control for the operations of the entire company : no individual budget can be out of line with it. This does not mean that each individual budget must show the identical percentage of profit, or return on investment, or balance between income and expense that the master budget shows. It simply means that all of the figures of the individual budgets, when added together, must equal the figures in the master budget.

Finally, if a budget is to be a useful guide and tool, it must be flexible. Budgets are based on forecasts, and forecasts, no matter how carefully made, are subject to error. Therefore, budget figures should be planned to allow for a certain amount of leeway. Each manager who administers a budget should review that budget regularly and recognize the fact that it will need occasional adjustment and bringing up to date.

Chapter 6
Developing a Marketing Plan

Now that facts have been gathered and the data analyzed, the company's problem is to use this information to decide how it will market its goods. The solution it devises is known as its marketing plan. The marketing plan charts the course the company is to travel.

A marketing plan is the result of an examination of the company's resources, its goals and objectives, and the environment and market in which it must operate. It is a corporation plan; but, like all other company plans, it cannot work in isolation. Each department within the company has its own plans, which at any given time are at different stages of completion. A successful marketing plan must take all these departmental plans into consideration. Thus a marketing plan is a consolidation of many smaller plans, each modified to mesh smoothly into the central plan and achieve the general policies and objectives of the company.

In developing a marketing plan, a company must make a number of decisions: What marketing objectives does it hope to achieve? What resources will it require? How can it budget for those resources? But its first step must be to decide how long it will use this particular plan.

TYPES OF MARKETING PLANS

There are two types of marketing plans: long-range and short-range. Short-range plans usually cover only the upcoming budget year; they are often called annual plans. Long-range plans generally cover a period of three or five years. Even longer-range plans—perhaps for 15 or 20 years—occur in some corporations, but they are seldom marketing plans. Marketers know that plans covering such long periods of time would be unworkable because of product innovations and constantly shifting consumer demand.

The purpose of these plans is to fit the company into the projected economy of the nation. Long-range plans are not static. If unforeseen changes occur in the marketplace, plans are altered to fit the new conditions.

Because it is difficult to make long-range projections with accuracy, long-range plans are more general than shorter-range plans. Their main use is in the establishment of organizational and financial policies which take considerable time to accomplish. Short-range plans, on the other hand, can incorporate considerably more detail than even the three- or five-year plan. As operating plans, their function is to achieve policies of a certain period of the long-term plan. Short-range and long-range plans are thus correlated.

Long-Range Plans

Long-range plans set a general framework for the next several years. An analysis of the data has given the marketer a handle on forecasting economic conditions, shifts in markets, and anticipated trends in consumer buying behavior. The data has also supplied estimates of trends in the sales of various company products, anticipated new product lines, and broad statements concerning company objectives. With this information at his disposal, the marketer is ready to plan the company's actions over the next few years.

It is only during the long-range planning period that a company has a chance to make decisions that will affect its destiny, for such decisions take time to accomplish. Long-range planning establishes priorities, while short-term plans implement them. If, for example, a company wishes to develop a new product, it must plan how to market it several years in advance so that funds and facilities can be provided for. If the decision is made in enough time, the budget, personnel, and other departments of the company can, in their annual plans, provide for the extra money, manpower, and other necessities. Small firms need the lead time even more than large ones, for they can little afford the risks of hasty decisions and usually do not have excess capital available for expansion on short notice. Long-range planning thus enables a company to decide what its goals are, a necessary prerequisite to accomplishing them.

Short-Range Plans

Short-range, or annual, plans are typically longer and more detailed than long-range plans. Long-range plans are broad in scope, while short-range plans map out strategy and schedules. They spell out budget allocations, priorities, target objectives, and dates for completion of each phase of activity. They emphasize specific products and their immediate production, promotion, and sales objectives. They contain department-by-department data, objectives, and goals.

A major purpose of short-term plans is to allocate the expenditure of funds and other resources. Most items in the short-term plan have both target dates and dollar figures assigned to them. In many cases, these are accompanied by a detailed breakdown of alternative procedures to be used if a strategy has to be altered in midstream. The short-term plan thus represents the practical application of the company's long-range planning.

SETTING MARKET OBJECTIVES

Marketing objectives are based on company objectives. Before a firm starts to devise a marketing plan, it must determine the demand for the product. Once the demand has been analyzed, it must be related to company policy. The company's general objectives and the budget it has allocated to marketing the product must be considered.

Marketing objectives indicate what the firm wishes to accomplish during a particular period in terms of sales, market share, reaction to competitors, and long-range goals. A typical statement of marketing objectives would probably include projected sales volume by product, market area, and specific market target; a statement of pricing objectives; a statement of profit objectives for each product and subproduct; competitive objectives by market and strength of competitor; and any other objectives peculiar to a particular industry. If the organization states all these objectives clearly, all levels of management can understand the firm's goals and work toward their achievement.

The framework of a marketing plan must be firmly based on reality. Marketing objectives must be reasonable and achievable. The major aid to setting realistic goals is the demand analysis, which predicts both the potential demand for a product and the company's potential share of the total market; model building, which permits each of the variables to be observed; and selection of alternatives based on appropriate decisions.

DEMAND ANALYSIS

The purpose of a demand analysis is to determine which products the company can sell and at what price. First the analyst surveys the entire market and predicts the demand for the product. Then he analyzes his competitors' probable reactions. This step is very important, especially to the smaller firm. When a small company develops a new product, it can be wiped out if a larger company decides to move in on the field, unless it has a patent on the product.

The demand analysis can be used with totally new products, with newly refined products, with old products in new packages, or with products having some other new feature. In each case, the purpose of the analysis is to anticipate the share of the particular market which the product can expect to receive.

Estimation of Demand. Most products are similar in some ways to existing products of other firms. Such products can be of assistance in determining how many products of which type can be sold in the market by this firm. Demand for most products is related to shifts in income at the consumer level, changes of prices of related products, changes in consumer attitude, as well as the impact of any changes of any of the above on the demand. To a large degree, each of these is measurable.

Generally, products are said to be either relatively *elastic* or relatively *inelastic*. The more stable a product's sales are, the more it is said to be inelastic. In other words, the more consistently customers purchase a product irrespective of its price, the more the marketer can predict its future demand. Staple items, emergency items, and certain prestige items fall into this category. At the opposite end of the scale are the products whose sales vary greatly with shifts in price. Such items are substantially dependent on changing economic conditions as predicators of sales.

The probable number of items that can be sold of most products is relatively predictable over time, as is the probable impact of any change in economic conditions on the item. It is important for a marketer to analyze the relationship of his product to other products as an aid in this regard.

Sales of large cars are affected by the price of gasoline, the general economic situation, and the general indebtedness of the buying public. Furthermore, it is generally recognized that if steel prices rise substantially, forcing auto costs up, certain car models will be priced out of the range of many former customers.

A demand analysis only indicates possibilities, not the final solution. It might, for example, show that several products have equal chances of success. It is at this point that decision making comes in. Resources are always limited, and the marketer cannot pursue every possible course of action. He must decide between alternatives.

TECHNIQUES OF DECISION MAKING

There is a mistaken notion among some managers that marketers feed information into computers and then, solely on the basis of mathematical computations, make decisions of whether and how to market a product. Mathematical computations do aid in making decisions, but the decision making process requires more than this. It is a skill that involves determining alternatives, comparing their probable outcomes, and deciding on one path. Factual data, experience, and intuition each play a part in the process.

The best way to make a decision is to define the problem, determine its underlying causes, develop alternative modes of action, and select the best alternative. The better the issue is defined, the stronger the possibility for making the correct decision. In considering which of two alternatives to choose, a company should look not only at the potential profit involved, but also at what indirect effect the decision might have on the firm. Can a product, for instance, assist the firm in entering a new market later on? Does it give the firm prestige which is valuable in other ways? Would the product's cancellation or the reduction of its production have adverse effects on other segments of the company's position—might it, for example, leave production capability unused? These are only a few of the questions that must be asked while making a decision.

Types of Decisions

Basically, there are two types of decisions: repetitive and one-time. Repetitive decisions continue to occur on a relatively frequent basis. They can be handled routinely once a decision process has been established. Because the alternatives are static for a time, the decision does not have to be rethought—it becomes a procedure to be used until the alternatives, the environment of the company, the nature of the marketplace, or other factors change.

The ordering of office supplies is a simple example of a routine decision. If a company has found that it uses an average of four packages of typing paper a day and that it takes three weeks to receive a new order of the paper, it will simply decide to reorder when the supply on hand is down to 60 packages (three weeks, or 15 business days, times four). It does not have to rethink this decision every time it runs out of paper.

Even more complex situtations can be readily solved once they have been carefully studied. If a firm has enjoyed a 20 percent share of the staple-product market over a period of years and other firms are not introducing innovations, it is simple to decide market potential for the product for the next year or two. If conditions were to change—if the natural foods industry were to take over a larger share of the market, for example—then the decision would no longer be a repetitive one; it would then become necessary to establish a whole new decision making process.

One-time, or nonrecurring, decisions are a different matter. They involve problems that have not come up before, at least not in exactly the same context. Therefore the firm must go through the whole decision making process—data gathering, risk estimates, gain projection, and review of how the alternatives fit into company goals and policies.

Many approaches to decision making have been developed; we shall consider the methods commonly used by managers.

Many of these techniques were devised for large firms, but small companies can often use them in modified form. In any case, awareness of alternative approaches should prove helpful to firms of any size.

Models

A model is a miniature or replica of an issue under study. Architects construct small-scale models of proposed buildings. TV weather reporters use models in the form of charts of the weather pattern. Businessmen use models to help in their planning. A model of the economy for the past year can be shown on charts and graphs. A model of the growth of sales of a product over the past five or ten years can be shown on a single chart. The intent in each case is to present a condition in perspective, so the viewer can gain insight into the total situation.

Managers use models to study all the possible alternatives before arriving at a decision. Models are safeguards. They help ensure that no alternative is being overlooked. Model-building may be considered a prerequisite to decision making. In the following discussion, we shall consider several of the most common models for business planning: role playing, computerized simulation, and game theory. All these models are simulations. Simulations are used in business to artificially duplicate the reactions of competitors and the marketplace in general to a given product and its marketing approach. They attempt to simulate situations and the interplay among the people involved in them.

Role-playing. Staff members can be used to participate by acting out various roles, pretending to represent the company's rivals. They are given as much data as possible about their role company, including information about its resources, personnel, and so on. Actors do not have to react in any set way to an event. For example, if the simulation is designed to determine how competitors might react to a strong promotional effort, participants will decide on a course of action based on their own personnel opinion of the best way to solve the problem.

Although at first glance the role-playing approach might seem too simplistic to be useful, it enables the company to evaluate a plan from all angles and alternatives. At the same time, staff members gain valuable experience in making decisions, even if it involves theoretical situations. Role-playing is also used with considerable success in personnel training situations. It has been shown that once people begin to respond emotionally to a hypothetical situation, they carry out the role they are playing as though the circumstances were real.

Computerized Simulation. Computerized simulation is a more sophisticated form of the role-playing approach. Here the computer plays all the roles. Pertinent data is fed into the computer and then programmed to compute certain business actions under a number of variables—changing business conditions, company pricing changes, new product introductions, and so on. The data may be very complex, but the computer can project the reactions of competition and the marketplace for the many different courses of action which the company is considering. The advantage of using the computer for this task is its unique ability to handle such complex instructions quickly. This permits a more detailed set of planning variables to be included in the experiment. The success or failure of such simulation, of course, depends upon the accuracy and completeness of the data fed into the computer.

Small companies can rarely afford the expense of renting a computer and hiring personnel to prepare a specific computerized simulation for themselves. However, they can sometimes use part of the data obtained from such models by other firms. Members of some trade associations can make use of industry models, which at least give more detail concerning the probable situation within the industry in general. Companies can use this information to project their own alternatives. This would be a help in studying the probable impact certain types of decisions by other firms would have on their own actions.

Game Theory. Game theory is another model for predicting how competitors will react to a company's marketing plan. Unlike role-playing, however, it is based solely on historical data. The participants in the game are not allowed free rein in their reactions. Game theory argues that the history of how competitors have reacted to similar situations in the past is a good indication of how they will react to future conditions. If past data shows that for every time a given company has placed a full-page advertisement in the local newspaper a competitor has placed a similar ad, chances are that it would react in the same way if the first company were to place an ad again. This type of information is useful not only to these two companies, but to the entire industry. The confrontation forces a chain reaction. Small firms can begin to predict within what price ranges they can act with relatively little risk and at what point larger firms will feel pressed and respond.

Like role-playing, the game theory simulation can be computerized. A vast amount of historical data is fed into a computer, which is programmed to predict reaction to alternative plans according to historical precedent. For example, the following postulates have grown out of game theory.

1. The more aggressive the action of one firm, the more aggressive will be the response from that firm affected by the action.
2. The more independently a firm acts with regard to its industry, the more the probability that other firms will respond aggressively when the firm finds itself in economic difficulty, even if the difficulty is only temporary.
3. The smaller the firm, the wider its disparity from the rest of the industry.
4. Generally, the more aggressively a strong firm attempts to injure a weaker firm, the stronger will be the attempt of the remainder of the industry to respond to the weaker firm's defense.
5. The fewer the number of competitors in an industry, the greater the risk to a firm that acts independently.
6. An abrupt change in the competitive nature of an industry provides the safest time for independent action.

These and other considerations can be, and are, reduced to formulas that test the probability of actions based on prestated conditions. Small firms can make use of these principles. They can also keep track of competitors' reactions under various economic and competitive circumstances. This will enable them to predict their competitors' future actions with a fair degree of accuracy.

DECISION THEORY

We have analyzed the demand for the product and have discovered an approach to methodically reducing complex problems to a point where they can be studied for decision making. Decision theory is not a theory in the strictest sense, but is rather a method of reaching decisions. One particularly useful method for decision making is probability theory.

Probability Theory

After managers have developed models of what their competition is likely to do, they are ready to weigh the probable outcomes of their decisions. This approach is known as *probability theory*.

The general principle of probability theory requires that each probability be assigned a specific value. Perhaps a firm is wondering whether or not a competitor will introduce a product onto the market during the next year. After looking at the available data, the planners may assign a probability of 60 percent to that occurrence. If the event does happen, a probability might be placed on the loss of sales of our firm's product. It might be determined that, based on the competitor's last introduction of a new product, there is an 80 percent chance that our sales will be cut by 10 percent; but a probability of only 20 percent that they will be cut as much as 25 percent. These probabilities, along with many other variables, can be shown on a "decision matrix," or chart of risks and alternatives.

Decision Criteria

To a large extent, decision making depends on the manager's own attitudes. No matter how carefully a company has studied the problem, competitors' actions and the precise nature of the economy remain uncertain to some extent. The manager may therefore either take an optimistic or take a pessimistic view of how things will go. In operations research, the terms for these alternative views are *maximax* or *maximin*. A third variation is called *minimax*.

Under the criterion of optimism, or maximax, the decision maker looks at the bright side of things. The assumption is that rivals will be unsuccessful in their competition and that the economy will hold firm. Therefore, the decision maker will decide to produce or stock large quantities of products and to place a high sales price on them. If correct, profits will be maximaxed. On the other hand, if the company's position or the economy has been misjudged, it will suffer the greatest possible loss. Not only will the company have overproduced or overpurchased the merchandise, but because of the high prices placed on it, competitors' products will appear more attractive to the cost-conscious customers of a depressed market. The criterion of optimism leads to the most extreme position for the decision maker. Chances of potential profits and of potential risks are maximized. The company will either be very well off or will be bankrupt. This, then, is the nature of aggressive, optimistic planning, or maximax.

Under the criterion of pessimism, or maximin, the assumption is that all decisions will go against the company. Even if indications are that the competition will not introduce a new product and that the economic conditions will remain strong during the planning period, caution is the byword. Here, the decision maker permits only a moderate supply of products to be purchased or produced. Goods are sold at moderate to low prices, the goal being a minimal profit over the risk of not moving any merchandise at all. Besides, according to this line of reasoning, at this price customers might be attracted away from competitors.

Under this pessimistic, or maximin, approach, the firm will yield a moderate profit if conditions are good. But it will have lost the chance to make really large profits—a firm operating under the maximax approach will have preempted them. However, under adverse circumstances such as a recession or unexpected competition, the firm will still receive a modest profit; and it will be far better off then it would have been had it attempted to operate under conditions of optimism which, in fact, did not develop.

Some managers, unable to choose between maximax and maximin approaches, employ the criterion of regret, or *minimax*. This is not another approach to decision making; it is a result of fear of choosing either pole of approach. The marketer decides on compromise, which will most likely cause him a minimum of regret. Of course, it will never lead him to the heights of elation, either. In planning, compromise is important—but companies that wish to grow powerful must take some bold action conversely. If management is afraid to make effective risk-taking decisions, the firm will jeopardize its growth potential.

One other approach to this type of decision making deserves to be noted: the La Place criterion, or criterion of rationality, which is based on estimates of probability of happenings. This criterion is very useful if the appropriate information is available, but is restricted to large firms because of the complexity of the criterion. A look at Table 3 will give you an idea of the criteria.

Decision Trees

It is important for the marketer to visualize what the outcome of alternative decisions might be. The decision tree is a useful model. It is pictorally shown as a tree lying on its side with its branches leading to alternative courses of action as well as the potential results of each alternative under both favorable and unfavorable circumstances. It can be used as an alternative to a matrix that pays off. It consists of (1) decision points, or points at which decisions must be made, (2) uncontrollable events which will have an effect on the outcome, and (3) results of alternatives under the various conditions. These elements are presented in a diagram that resembles a tree whose limbs and branches represent courses of action. Figure 3 shows a simple model expressed in the form of a decision tree. Here, the decision point is shown as a square. Uncontrollable events are indicated by circles. Any other symbols will

Table 3. Decision criteria: Net gain or loss for a product over a year.

ECONOMIC CONDITIONS	OPTIMISM (MAXIMAX) $	PESSIMISM (MAXIMIN) $	MINIMUM REGRET (MINIMAX) $
Favorable	80,000	20,000	30,000
Mild	20,000	15,000	16,000
Unfavorable	(30,000)	12,000	5,000

Figure 3. A simple decision tree.

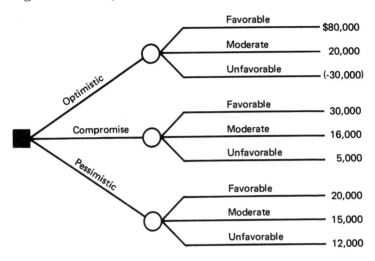

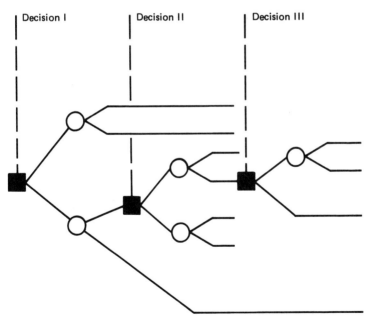

Figure 4. A more complex decision tree.

serve as well to make the point. This type of visualization enables the firm to diagnose a situation quickly and accurately.

These diagrams are usually more complicated, since a marketing plan has to incorporate many decisions, each of which operates in interaction with other decisions. Sometimes a subsequent decision can necessitate changes in a preliminary decision, a concept known as *rolling back*. Therefore, the marketer should think through all the consequences of a decision before settling on it. This is where a more complex decision tree comes in. It points out the various possible decisions and their probable effects on each other. This type of decision tree is shown in Figure 4. Its purpose is to force the manager to analyze the consequences of a particular decision, to think through the entire interlinking chain of decisions to their eventual effect on the success or failure of an objective.

IDENTIFYING THE POSITION OF FINANCIAL RESOURCES

Once objectives and specific product plans have been established, the types of company resources needed to implement a specific strategy must be determined and assessed. Money, personnel, and production resources must all be considered. The biggest danger here is that the firm will overcommit its resources, leaving no room for later adjustment and modification of the plan. This issue will be discussed in the next two chapters. At this point it is important to recognize that planning cannot take place without some consideration of resource allocation. Allocation models can help the marketer in this task.

Allocation Models

Allocation models are intended to help the marketer determine how to use his limited resources for the firm's maximum benefit. They plot the allocation of funds or other resources in various ways, showing what would happen if the resources were used in any of these ways. Allocation models serve two purposes. First, they indicate how resources should be channeled to best effect a specific marketing plan. Second, they clarify the financial soundness of each allocation under consideration. Thus the marketer can see how to allocate his resources to eliminate as many risks of financial failure as possible and to gain the greatest return on investment. Allocation models are a great help in budget planning.

BUDGETS

Planning requires facts. In business, facts are typically presented in unit or dollar terms. This information is usually provided in budgetary form. Marketing is only one function of a firm. Marketing needs must be assessed and placed into a relationship with the needs and resource allocations of other departments. The budgeting system operates at two distinct yet complementary levels—company-wide and operating unit. In the company-wide budget, resources are estimated and needs projected for the company as a whole. With operating unit budgets, each unit or department of the company prepares its own budget, compares it with total company needs, presents its case to management, and then, if the total company budget does not allow the anticipated amount of support, the unit must adjust its figures. Like other departments, marketing must base its programs on total company needs and limit itself to the resources available for marketing during the planning period.

Types of Budgets

For a large variety of reasons, a specific company may have any number of budgets. Budgets assist not only in planning but also in implementation, scheduling, and control. Our concern here is with budgets only as a part of the planning sequence. Corporate planning budgets usually begin with statements of (1) projected income and expenses, (2) projected financial condition, (3) projected cash flow, and (4) projected capital expenditures.

Statement of Projected Income and Expenses. An analysis and statement of income from sales is basic to the planning process. Forecasts of sales, based on the economy and competitor's actions, come early in the planning process. In the budget, these prospective sales figures are organized by product and territory, and they provide a sound basis for estimating the proposed income for the planning period. This income figure is then balanced with detailed expense figures showing cost of goods sold, expense of operations, and other appropriate expense categories. The statement must be clear enough to enable planners to see the relationship of sales to expense items.

Statement of Projected Financial Condition. The primary function of this budget statement is to indicate where the tightest financial resource problems will be. Projected monthly or quarterly balance sheets can serve this purpose. Knowing where the crunch will come permits special attention to the crucial periods of the planning period.

Statement of Projected Cash Flow. Cash flow statements are simply estimates of cash use requirements over the planning period. They take data from the other budgets and place it in a format which permits management to observe the company's current position.

Before any revenue is received from a particular product, raw and processed materials for its manufacture must be purchased, advertising paid for, and sales calls made. During this period the firm must have enough money on hand to proceed with the next stage. The cash flow statement is intended to tell the company how much money it has on hand by projecting expenses and receipts during each month. It allows management to see how much money is needed for how long to keep the company from running into trouble. It also permits management to note if projections are not being fulfilled. Such a case will require quick action from management to adjust the plan and strategy.

The cash flow statement is perhaps the most vital of the budgets. It focuses on what top management and marketing management need to know. It also represents one of the easiest ways for management to determine (1) which projects it is practical to propose for the planning period, and (2) which should be dropped because the projection is in terms of the actual cash needed during each period to fulfill the company's goals.

Statement of Projected Capital Expenditures. A company may decide to invest money in capital expenditures such as building improvement, new equipment, and other one-time improvement costs. The capital expenditure statement presents the costs and anticipated returns of such proposed expenses in dollar amounts.

Budgets and Marketing Planning

In preparing their departmental budget, the marketing people must estimate sales and indicate profit probability. Their statements will be reviewed by others in the firm, and a total picture will be developed for the entire company. Then the marketing department will be given expenditure estimates for the planning period. At this point the marketing planners must adjust their budgets to match the amount provided in the master budget. Of course, they may discuss problems with top management and make some adjustments in the amount allocated for marketing. But the marketing budget is determined by the total company picture, and marketing planning must be adjusted to fit into the total picture. The typical marketing budget includes the following specific items and sub-budgets:

1. *Sales Budget.* The sales budget lists sales by product, prices, territory, and estimated returns.

2. *Inventory and Materials Budget.* This is a statement of the inventory necessary to achieve the sales levels anticipated for a particular period. It includes estimates of materials and supplies needed to fill orders promptly.

3. *Cost of Goods Sold Estimate.* This statement includes material costs, labor costs, and manufacturing expenses. The costs of goods sold is probably the most used estimator, and usually exists before the rest of the budget is prepared.

4. *Sales and Administrative Expense Estimates.* This category could be broken down into two separate budgets, for it covers independent categories: sales and sales support expenses, such as wages and pro-

motional expenditures, and administrative costs of the marketing area. These costs tend to be high, but they can be overlooked if one is not careful. Many companies have found that they did not provide enough funds for proper promotion, and as a consequence they are in the unfortunate position of having made substantial commitment and sacrifice in support of a product without receiving the anticipated benefits of sales. Perhaps worse, they have lost the chance to spend the money on other profit-producing alternatives.

5. *Research and Product Development Budget.* Research is the key to future growth and the supplier of feedback on current operations. Budgets for marketing and product research should be large enough to permit these departments to fulfill their functions.

Planning each of the above budgets requires considerable effort, but it is worth it if they provide the data to allow appropriate use of resources. However, no matter how good a budget is, it cannot foresee every contingency. Therefore all budgets should allow for funds for emergency use. Such funds provide not only an insurance against calamity, but also the flexibility necessary to take advantage of unexpected opportunities. What the precise level of leeway in a budget should be depends on the industry and corporate situation.

THE PLANNING AUDIT

An audit typically reviews the company's accounting to determine whether the firm has functioned as required by law. What is expected as a result of an audit is some substantial degree of assurance that funds have been expended honestly and appropriately.

A company may also decide to conduct an audit to see whether its expenditures have been related to its anticipated goals and objectives. This sort of review is known as a *planning audit*. Like the other type of audit, it is a check— this time, a check on how the plan is developing to date.

Establishing a Planning Audit

A planning audit checks predetermined benchmarks or checkpoints against performance up to a certain time—say three months. Typical benchmarks are statements of sales performance by product after a period of time; the ratio of sales to industry sales; advertising expenditures; and the cost of goods sold.

The audit permits a comparison between the planned goal and the achieved or actual goal for the period. If projections led the firm to expect a specific product line to sell at the rate of 100,000 units in the first three months, while the audit figure shows that actual sales amounted to only 60,000 units, there is cause for concern; the audit has proved very helpful, because now management can adapt its plans to fit the altered situation. If the planned and the actual figures come close to agreement, the plan is reinforced, and management may turn its attention to other problems. Planning audits, particularly those that deal with marketing areas, are not perfect, but they are helpful in determining the present state of the plan. The procedure is easy enough to be practical for all marketers.

PREPARING THE SCHEDULES

In order to be useful, a plan should be broken down into schedules. The timing of each element must be spelled out. Every expenditure must be calculated.

Once the marketing plan has been structured in this way, it can serve as the guide for the planned period as well as the control.

Schedules are of many types. Timetables, cash flow statements, graphs, and charts are all important schedules, and there are more sophisticated techniques which in some cases provide even more detailed and better information. The following discussion will deal with several of the more common approaches to scheduling.

Timetable of Deadlines

Deadlines are very important to scheduling. They are predetermined points by which certain goals must be reached, or at which certain actions are to take place. Let us consider briefly two types of deadlines.

Data Deadlines. Data deadlines are used for audit, adjustment, and control of the plan. They set specified times for collecting data. Many types of data may be gathered on a deadline basis, including monthly sales records, quarterly income and expense statements, inventory statements, sales expense statements, quarterly balance sheets, and economic and competitive-industry data.

The purpose of the deadline in each case is to remind management of how its plan is progressing. This in turn should result in a restudy of the plan, if it is shown to be faulty. Deadlines tend to make the members of the management team strive to achieve their projected objectives prior to the due date.

Action Deadlines. Action deadlines define the dates by which some type of action is to take place. They are movement dates for some important move predetermined by management. Here are some of the typical action deadlines for marketing managers.

1. Physical events
   Shipping dates: Merchandise must be shipped by a certain date for some reason, perhaps to be on the shelves in time for Christmas shopping.
   Merchandise markdown dates: Perhaps management wants to take advantage of a special bargain day, like Columbus Day, or perhaps it has been determined that this is the best time to mark down merchandise to get the jump on the competition.
   Cancellation of production dates: Management has decided that if sales do not improve by this date, they will cancel production.
   Warehouse clearance dates: The space might be needed for new merchandise by this date.
2. Implementation events
   Advertising campaign dates: It has been determined that this period will be most effective for the campaign.
   Extension or cancellation of production dates: If the product is doing better than expected, production will be increased; if worse, it will be canceled.
   Dates to purchase under discount: This might be at the best time to purchase needed equipment.
   Alteration of plan implementation: Management has decided that this is the deadline for implementation of the original plan; if things aren't working out, alternative procedures must be adopted.

Each of the above deadlines involves specific actions or decisions by the firm, and to bypass any one of them inadvertently would impose an unexpected hardship on the company. Deadline timetables are particularly useful for small companies. Indeed, they alone may provide enough control for management to monitor its plan. For larger organizations, however, deadlines form only the first step in plan scheduling.

Project Flowcharts and Diagrams
Industrial manufacturers have traditionally used flowcharts for their projects. Flowcharts show at a glance both the planned progress of an undertaking and its actual state of completion. Figure 5 is a typical flowchart. The dates are placed along the bottom of the chart, while designations of the various projects go along the left-hand side. The planned events are marked on the ribbon going from left to right. Thus, many items can be shown on one chart, and many activi-

Figure 5. A project flowchart.

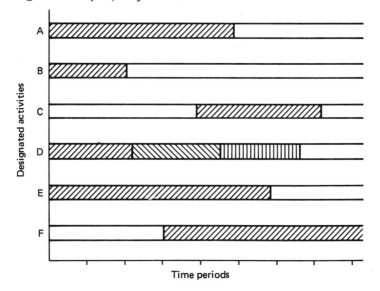

ties can be followed. Advertising campaigns, product development projects, and sales campaigns are but a few of the types of planned marketing activities that can be charted in this manner.

Although flowcharts are an excellent means of following a particular project, each new project must be charted individually, which entails using separate pages, or artists' boards, for each. Then too, each project is made up of a complex set of activities. Not only do they require considerable space, but they also force the viewer to look back and forth to compare the several portions of the project.

A further drawback of project charts is that there is no provision for comparison of relationships of one activity with another. In reality, many projects can proceed only so far without needing data or resources, such as parts or materials, from another project. A special notation or coding has to be made on the chart to show where such information can be found, and the viewer is forced to consult several charts for the information. This makes it difficult to spot relationships early or easily. Flowcharts, then, have an important role in simple situations and can be used to good advantage by small companies. But most do not provide the detail needed for more complex marketing plans.

**COMPLEX SCHEDULING TECHNIQUES:
PERT AND CPM**

One of the most effective approaches to future-directed control is the formalized technique of network planning. Because of the growing complexity of its project management, the government had scientists develop a scheduling system to present complex relationships clearly and in perspective on one sheet of paper or wall chart. After months of study, the team developed the PERT system and its variation, CPM.

PERT stands for program evaluation and review technique, and CPM means critical path method. The basic concept behind the techniques was to devise a chart that would arrange all activities and events so that they could be seen in relationship to each other. Thus, they would focus attention on critical areas before they ballooned into crisis situations, and actual accomplishments would not be missing desired goals, to the manager's frustration.

The PERT network shown in Figure 6 will illustrate and show how to identify the most critical path, or the one with the least slack. But first, what does the chart represent? The circled numbers are measurable, verifiable events, and the arrows are estimates of weeks required to complete an event. (The unit of time may be in days, weeks, months, even years.) All activities are tied to events. Each begins with one event and ends with another. An event may require the completion of more than one concurrent activity, but no activity can lead to more than one event.

Each event is numbered so that data can be traced easily, and each activity is referred to by the numbers

Figure 6. The basic PERT network.

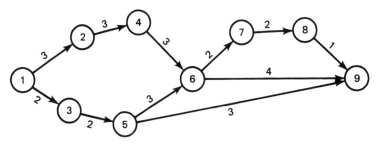

of its beginning and concluding events. Thus, in Figure 6 the activities at the left-hand side of the chart would be referred to as activities "1–2" at the top of the chart and as "1–3" at the bottom. The length of the lines bears no relationship to the length of time required to complete the activity. This simplified PERT network shows the ease with which the PERT chart can illustrate the way programs interlock.

Activity Completion-Time Requirements
The PERT network can show each activity in terms of the number of weeks required to complete it. The time is determined by the project manager, who provides three types of information, determined on the basis of previous experience: (1) the quickest, or shortest, amount of time the activity can reasonably require; (2) the slowest, or longest, time needed if things go badly; and (3) the probability of either taking place. From this information, the average probable time is derived. It is placed directly on the network for all to see.

Finding the Critical Path
Once all the time estimates are placed on the network, the planner can add them up along the activity lines and quickly determine how long it will take to complete the entire sequence of events from beginning to end, if all goes as planned. But the manager can also gain another, even more important, computable piece of information called the *critical path*.

Look closely at the activities in Figure 6. You will note that it takes less time to complete some of the activities that lead to a particular event than it does for others. For example, to reach event 6 by taking the 1–3–5–6 route along the bottom of the network takes up 7 weeks. If you start from the same point, but go across the top and down, 1–2–4–6 takes 9 weeks. This tells us that nothing particularly damaging will happen to our plans if anything should go wrong along the lower path, for there are a full 2 weeks of leeway in the sequence. But if we cross the top, it is a different matter. Let's say event 2 is 10 days late. This delay will hold up the entire project unless something is done right away. Event 6 cannot take place until all the activities preceding it have been completed. The top, then, is the critical path of events. There is no slack time, for everything will be held up if the planners do not follow it carefully.

Costing the PERT Network

PERT networks can be constructed to show more than estimates of time. They can reflect cost estimates for each activity as well. These cost estimates can be provided both for the longest and the shortest time periods. They help the planner determine which, if any, activities along the critical path should be given a reduced time schedule to help pick up time for the entire project. This lessens the risk of losing the schedule, which can be very expensive.

Unsophisticated PERT Networks

A small enterprise can establish a relatively simple PERT network. Nothing more than patience is required to follow these steps.

1. Separate all activities into their smallest segments. Since PERT is actually scheduling by detail, the first step is to identify each of the activities and events that take place in a planning process. Begin by taking the deadlines that have been established, the

cash flow statement, the dates for receipt of merchandise, and all other events that come to mind. Then see if you cannot break each of them down into further subactivities. Continue to subdivide until you are sure you have reached the smallest point.

2. Determine time and cost estimates for each activity. Estimate the average time, but make sure it represents the most probable time for each activity. Determine the cost estimates for each activity for the fastest and the slowest periods, but do not attempt to determine which cost estimates to use until the critical path has been determined. Cost estimates should include *all* costs—overtime, shipping costs, and so forth.

3. List and number the events. The number for each event will become its code or reference number. Do not be concerned about matching the sequence of numbers to the sequence over time, that is, don't worry if the event you call "14" ends up ahead of event "12," since it is important only as a code.

4. Determine the sequence of action, the exact relationship of activities to events. It is not uncommon for three separate activities to have to be completed in order to achieve an event. This relationship must be observed and noted before the network can be constructed. It is also necessary to establish the proper sequence of all events. If you have inadvertently listed an event more than once, assign only one number to it. Use the extra numbers for later events.

5. Construct the PERT network. At this point you are ready to place the information on the network in the form of nodes (circles) and lines (linking arrows). Since you might have to change your mind, use a bulletin board or small magnetized nodes that can be shifted easily.

6. Determine the critical path. Now you can study the network to determine where shorter activity might be beneficial, even at higher costs. You can then look for areas in which certain activities reach the event stage ahead of other activities that lead to the same event. In such cases, you might decide to select longer activity times if it will mean lower costs or will reduce the pressure on other activities. Such adjustments do not have an adverse effect on the outcome of the project as a whole.

The work entailed in these steps demands accuracy, and it is time-consuming the first time around. But your next networks will be far easier to achieve, and you will have a valuable tool at your disposal.

WATCHING OUT FOR PLANNING PROBLEMS

Market planning requires considerable effort and attention. Every stage presents opportunities for distortion and misestimation. Here are some pointers that can be used to lessen the risks of error in preparing the plan:

1. Establish an outline of the plan before attempting to fill in the data.
2. Check out sources or outside information used in the plan—economic reports, competitive performance, and so on.
3. Bring data up to date. Assumptions should be based on as current data as possible.
4. Base your company estimates on as many related facts as the risk necessitates. Do not search for complete data on small-risk or small-investment categories.

5. Avoid overplanning. Try to spend only a limited amount of time on the planning process, and alter it only when substantial new information requires it.
6. Involve those who will be carrying out the plan. A management-only plan will not be the strongest one for the company.
7. Avoid committee decisions. They tend to be compromises, not decisions based on hard data. Assigning responsibility for specific data will bypass this problem.
8. Provide for adjustment at all stages of the plan. Avoid a plan so specific that it cannot be adjusted without great effort.
9. Establish a system for feedback and monitoring of the plan.

The marketer who stays on top of these details will avoid many of the pitfalls in planning.

PART THREE
STRATEGY

Chapter 7
Orchestrating the Marketing Dynamics

Managerial planning involves the use of money, manpower, and materials—what ways, how much, and when. The manager has three tasks: to collect and analyze the company, the product, and the market; to home in on the precise market and operating environment and try to forecast whatever opportunities and risks lay in store; and finally, to combine all this information into a solid marketing plan, armed with a strategy that will help it succeed.

The marketing plan is a cohesive structure. It is made up of a vast assortment of assumptions, facts, and figures, all organized in an orderly manner. The plan is designed to reflect company policy and to achieve company objectives. It gives the company the information it needs to determine tomorrow's goals and the direction company progress should take. In a way, the

Note: Much of the material in this chapter is based on notes and lectures, as well as a new monograph by Norton Paley, *Solving Marketing Problems: A Practical Guide to Marketing Strategy* (Alexander-Norton, New York, 1978).

marketing plan is the vehicle that takes the company from the present into the future.

However, the plan is simply facts and figures. Some sort of catalyst is needed to convert the raw data into a strategic plan. A plan without strategy is like a group of isolated activities that lack a unifying force. Strategy is the art of coordinating the means (money, manpower, materials) to achieve the end (profit, customer satisfaction, growth) as defined by company policy and objectives.

Companies do not compete against other companies, nor do products compete with products. Managements are measured against managements, and strategies compete against strategies. Business is a contest of one manager's ability pitted against another's, and the mark of the successful manager is his or her ability not only to select the right facts and figures for a marketing plan, but also to weld them together with the most profitable strategy.

Business strategy has its roots in military strategy, and much of what has been developed over the 5,000 years of military history has its parallels in business. A Byzantine general, Belisarius, who lived during the first half of the sixth century, said that the most complete and happy victory is to compel one's enemy to give up his purpose, while suffering no harm oneself. Although today's businessman would phrase it differently, it is exactly what a manufacturer wants to do when his product's share of the market is threatened, and what a shop owner wants to do when faced by the competitive pressure of a new discount center opening in the area.

Whether the strategy is military or business, the goal remains the same—to achieve an objective, as specified by company policy, by employing human and material resources against a variety of obstacles. Developing a marketing strategy involves understanding the factors that make up the environment in which it is formulated and in which it operates, and knowing which of these factors can be controlled and which cannot. It also involves recognizing and knowing how to use the characteristics most frequently found in successful strategy:

- Speed
- Indirect approach
- Unbalancing the competition
- Concentration of strength against weakness
- Alternate objectives

THE STRATEGY'S ENVIRONMENT

Strategy is formulated and carried out in an environment made up of a number of interacting business and social forces. Basically, these forces can be divided into two categories—those factors that a company can control and those over which it has little or no control. There are some factors that go beyond the individual manager's control, but which the company as a whole can handle. For instance, a sales promotion manager given the job of planning a campaign for a new product probably has no input into the way that product is designed, and little control over the funds he or she can use, but because other members of the staff manage and influence these variables, they can be considered "controllable."

Controllable Factors
Those elements of the marketing environment that are generally within a company's ability to change,

modify, or influence are called *controllable factors*. They include:

- Product or service (including all activities involved in design and development)
- Price
- Promotion (including such activities as personal selling, advertising, display)
- Distribution (including selection of distribution channels and means of physical distribution)

These factors—the product or service and the various means used by a company to sell it to a particular market—combine to make what is commonly known as the *marketing mix*. The proper mix of these four components in a profitable strategy depends on the needs of the market to be served, the extent of company resources, the general condition of the economy, social and legal considerations, and the nature of the competition. Each element of the marketing mix presents a number of challenging variables for the marketing manager. Here are some of the techniques at the disposal of the manager. By definition, each element of the marketing mix can be controlled by means of strategy.

Product or Service Planning and Development. The actual item for sale, as well as all the activities that go into its development and manufacture, such as type and quality of materials; choice of brand, packaging, and labeling; and degree of standardization, comprise the product's involvement in the marketing mix. It can be a service as well as a tangible item. The company that markets maintenance service is just as concerned with its "product" as the company that sells furniture.

If a company manufactures the product it sells, it must plan everything from what combination of colors or styles or flavors should be marketed, through what shapes and sizes of containers should be used, to what brands and labels should be chosen. Decisions may also involve what services the firm should perform in support of its products, what guarantees the products should carry, and under what conditions returns or replacements are to be permitted. If the company is a wholesaler or retailer that buys products from manufacturers, decisions must be made as to which products the company should stock and what services should be offered to back up the product assortment.

Price Determination and Administration. Pricing a product is a challenging part of the marketing mix, even though the final statement in the form of a specific price may appear to be relatively simple. The right price has to take into consideration the competitive products, structure, and practices, as well as the condition of the economy.

At the manufacturer's level, such considerations include necessary credit terms, returned goods policy, delivery costs, warranties, guarantees, and discounts, not to mention production costs and a margin for profit. At the retail level, the pricing decision also takes into account the average markup, markdown, shrinkage, and similar costs, as well as legal considerations at all levels. The consumer's attitude must also be factored into the price. It is important to remember that customers view price both objectively and subjectively. Thus, a price that is just a little too high may turn potential customers away. On the other hand, the price should not be so low that even potential new sales will not make enough of a profit. It is a matter of striking the right balance. Manufacturers have to be careful to offer their products in such a way that they return an acceptable profit without turning customers away, and retailers must select product price lines that will attract customers to the store and result in a purchase.

Promotion. The art of promotion encompasses all those activities, both informative and persuasive, that communicate news about products to a selected audience. Sales promotion includes advertising, display, personal selling, publicity, and public relations. (Although public relations is often defined as a nonpromotional activity intended to build and maintain good community relations and a positive company image, the effort automatically has its promotional effect.)

Promotional decisions involve anything from choosing what types of media to advertise in to allocating money for dealer aids or special window displays. Manufacturers direct their promotional efforts both at the dealers who carry their products and at the customers who buy from those dealers. Retailers usually have a single promotional target—that all-important ultimate customer.

Physical Distribution and Channel Management. The product must be gotten from the manufacturer to the customer for the marketing process to be completed successfully. Historically, physical distribution has been handled on a piecemeal basis, which is grossly inefficient. But advancements in transportation and a systematic approach are beginning to eliminate the bottlenecks. Manufacturers are concerned with the channels of distribution, except in cases in which they sell directly to the ultimate customer. A distribution channel is an organized network of wholesalers, brokers, agents, and retailers that links producers with users. All marketers' physical distribution methods are to be used for such areas as traffic and transportation, warehousing, order processing, and some types of inventory control.

The Marketing Mix. The combination of these controllable components of the marketing mix is the crux of formulating a marketing strategy. A large enterprise that has many product–price–promotion–distribution combinations can have as many as 5,000 possible marketing mixes from which to develop a strategy. The manager who faces this kind of choice may very well find an answer in computer simulation models that project and compare the results of various types of strategy in a particular market situation.

However, although the computer, or more generally, electronic data processing, is eliminating much of the drudgery connected with working out and comparing the results of taking various kinds of action, the manager must exercise judgment in making the final decision. In spite of all the advances in management science techniques being used, marketing remains just as much an art as a science. Any field that attempts to theorize about people—what they need, what they want, what they will buy—requires a certain degree of intuition to make the leap from one empirical observation to another. The determination of why people react as they do, to say nothing of how they will react next, is still a very uncertain science at best. There is room for instinct and human intuition in determining marketing strategy. Today's educated hunch profits from the information automation can supply.

Uncontrollable Factors

Those elements of the environment that are generally beyond a company's ability to change, modify, or influence are called the *uncontrollable factors.* They include:

- Consumer attitudes and tastes
- Economic conditions
- Legal and social constraints (federal, state, and local legislation about trade practices, standards, prices, exports, and imports in particular)
- Available company resources
- The competition

Figure 7 depicts the interplay of marketing management variables, which *can* be controlled, with environmental variables, which cannot. Sometimes a specific industry or product can determine the degree of control—or lack of it—that a company has over one of these factors. It is very hard for a company that markets basic construction supplies, such as cement blocks or lumber, to influence consumer attitudes or tastes. Builders already know that they want to handle a specific construction job. On the other hand, a customer in the market for a new car might be swayed by particularly persuasive promotion. A man may start out hunting for what he considers a basic family car, and end up buying a roomy sports car.

A company may not be able to control certain conditions, but that factor can have a significant impact on the company. For example, an organization has little control over the economic conditions within its marketing area, but the economic situation will influence what people will buy and how much they will be willing to spend. Here are a few of the ways in which uncontrollable factors influence the type of marketing mix a company chooses.

Consumer Attitudes and Tastes. Consumers can be divided into different groups that underscore certain attitudes and tastes, according to the geographic area in which they live, average age, income level,

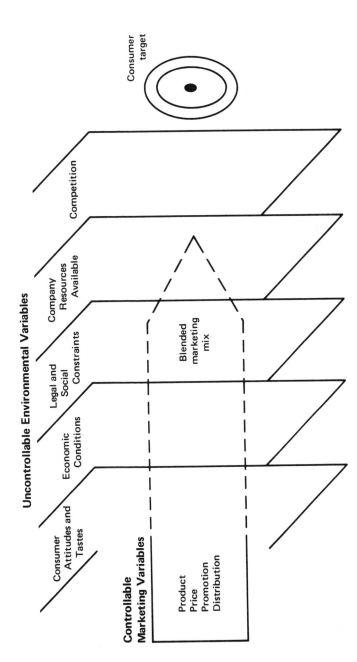

Figure 7. Elements of marketing management strategy.

educational background, and ethnic heritage, among other factors. A marketer has to recognize and understand the particular attitudes and lifestyles of the target group before attempting to serve them properly. Whereas those attitudes and tastes cannot be controlled, they will determine what the product assortment should contain, the price levels, what services should be offered, even how the outlet should be fixtured and what hours it should be open.

For instance, a dress shop in an upper middle class, middle income, suburban neighborhood would probably do best to concentrate on conservative styles at moderate prices, with a sprinkling of extreme styles at higher prices and only an occasional bargain buy. The store would probably be open during normal business hours and perhaps one or two nights a week; it should be comfortably fixtured; a good choice of services, such as credit arrangements and delivery, should be available.

Contrast this with a neighborhood delicatessen in an urban area that contains both moderately priced apartment complexes and high-priced cooperative apartment buildings. Such a store owner need not be too concerned with elaborate fixturing, as long as the food doesn't suffer. There should be a wide variety of deli items and delivery service if requested, and the store should be open 7 days and evenings a week if possible.

Economic Conditions. The economic environment of the market in which a company wants to do business is a vital consideration for that company. It is important to realize that the general economy is not the key factor, but rather the economic state of the company's potential customers. The relative prices of similar products on the market also play a role.

Marketers who supply all kinds of industrial equipment to steel producers do well when the steel industry is booming. Let the federal government cut back on its contract spending, or the automobile industry hint the possibility of a strike, and these suppliers will find their customers very wary about putting money into more equipment. The same is true of the consumer market. A retailer in a town in which unemployment is a problem and inflation is eating away at the paychecks of those who are employed would be foolish to bring in an assortment of new luxury products. He would do better to concentrate on finding and offering bargains to his customers.

Legal and Social Constraints. In recent years, business has had to grapple with regulations and social pressure. The legal constraints have occurred at the federal, state, and local levels of government. Government regulation is carried out in two forms—through laws already in existence that must be obeyed as long as they are on the books, and through recommendations made at various levels of government. These recommendations are frequently used by the Federal Trade Commission to suggest what businesses should and should not do. Although not actually law, such recommendations do have government clout behind them, which argues strongly for following them.

Today, there are laws and recommendations concerning many phases of operating a business, such as the methods of competition, prices, taxation, product standards, brands and labeling, and building specifications. As far as the marketer is concerned, it is sometimes helpful to remember that these legal restrictions have two goals: to protect the customer, and to protect the marketer against any unfair practices of another marketer.

The last few years have also evidenced a vast increase in pressures from various direct and indirect customer groups, thereby limiting or at least requiring a reevaluation of corporate behavior. Pressures for energy-efficient appliances, increased mileage per gallon in cars, pollution- and odor-free facilities, and fewer food additives, along with reactions to sales or component part purchases from certain countries for popular political reasons are some examples.

Available Company Resources. Planning must be handled within the framework of the available company resources: money, manpower, and materials. Managers must know the boundaries of their company's resources in order to work out a realistic plan. A successful marketing effort could very well mean more resources tomorrow, but the manager should not count on using them today.

A company should not plan to market a new product throughout the entire country if it lacks the funds to produce a sufficient quantity of the product or to promote it adequately. Money can be borrowed, and equipment and technical skill can be bought, but eventually the loan must be paid back with interest. That purchase is also an investment of present resources, and it must be judged to have a reasonable chance of returning a fair profit on the initial investment.

The Paradox of Competition. Although competition is usually treated as one of the uncontrollable factors in the market environment, there is a paradox in the fact that to change, modify, or influence competition is one of the most important goals of strategic planning. Nevertheless, managers do not have direct control over what the competition does or is. They do not dictate whether competing companies will change their pricing systems or their package designs. However, managers can and do try to alter the marketing situation through personal strategy designed to force the competition to react to their moves.

Competition is one of the key elements of resistance that must be overcome. Elaborate plans are made, products designed, prices determined, advertising worked out, salespeople briefed, distribution channels selected—in fact, the entire organization might shift into high gear often simply to outwit the competition. Thus, although an uncontrollable factor in one sense, competition is also one of the prime movers in strategic planning.

Meeting competition head-on through the introduction of a new product, or entering a new market, or even just attempting to maintain sales volumes of existing products in the face of increasing pressure should never be undertaken lightly or recklessly. A marketing strategy must be worked out that enables the plan to succeed without exhausting valuable company resources.

Before deciding on which strategy to use, a manager must use the facts he or she has collected to weigh the competitive situation in the market. Some markets should be avoided altogether, just as certain products should not even be considered, and certain competitors are best left unchallenged. The manager must be prudent in planning a strategy, and yet not be hesitant. The habit of blundering aimlessly into situations should be controlled, and yet if a good opportunity arises, the manager must act swiftly and decisively.

All the aspects of strategy hang together if the manager is creative and adept. Strategy is an intellectual exercise. It involves a mind-to-mind encounter between managers. The brain waves of strategic confrontation are ultimately transposed into tangible figures in the balance sheets of the competing companies.

THE TWO LEVELS OF STRATEGY

Strategy for a marketing plan is usually worked out on two levels—high-level strategy (or grand strategy) and lower-level strategy (or working strategy). To some extent, the division is also organizational. Top management plans and carries out the grand strategy, while middle management is often given the responsibility of planning and carrying out the working strategy.

Grand strategy is the overall plan for the way in which company resources are to be used to achieve company objectives. It works hand in hand with company policy, putting it into action. Grand strategy takes a long-range view about the company's objectives in the market.

A toy company, for example, might have a long-range plan to trade up its product offerings in order to increase the profit per unit without increasing production facilities. Its high-level strategy might specify that the bottom half of the price lines gradually be phased out over a period of three years, while new higher-priced items are added, one by one, as resources for their development become available and as testing in the market shows their probable acceptance. Here, the company's grand strategy is designed to take the company's products out of the discount outlets and put them into the higher-priced stores.

Grand strategy is worked out to meet the potential for future development and profitable growth at a minimum expenditure of the company's resources.

Part Three   Strategy

That toy company does not want to trade up simply by adding new, higher-priced lines, which would require more production expense as well as all the other costs of introducing new products. It intends to use its present facilities to phase out the lower-priced items, and then use the freed production time to turn out higher-priced items.

High-level strategy must be developed along the lines of the goal without wasting the company's strength. Self-depletion of resources is responsible for more cases of financial exhaustion of companies than competition achieves. There are numerous cases of companies that have overextended themselves, exhausted their resources, and collapsed, simply because they wanted the immediate satisfaction of winning a market, regardless of the cost and the long-range effect.

Working strategy provides a detailed plan that coordinates and activates every facet of the overall marketing plan. The working strategy may be developed by every manager responsible for handling a particular aspect of the overall plan. The working strategy is continually being tested and modified throughout the marketing effort.

Every manager in the toy company is involved in planning and carrying out a working strategy. The production manager plans how and when to retool the production line for each new product. The sales manager works out new schedules for his or her salesmen that would take them to increasingly higher-priced outlets. The promotion manager thinks up publicity schemes for each new product to underline its better quality.

The success of any strategy depends on sound calculations and careful coordination of the goal and the means to achieve it. The nature of the business world and the problems of predicting customer needs and wants make it difficult for even the most competent manager to make completely accurate calculations, but success is won by the manager whose strategy is based on calculations that come closest to that ideal.

SPEED: THE KEY TO STRATEGY

Perhaps the most important characteristic of any successful planning strategy is the ability to keep the action of the plan moving. Generally, if the action is slow, and the company engages in a prolonged, dragged-out marketing effort, it will not be able to withstand the drain on money, manpower, and materials. The result is likely to be failure—not only of the marketing effort, but sometimes of the entire enterprise. What if a competitor comes out with a product or promotional plan similar to the one your company has been struggling with?

A marketing campaign to launch a new product requires considerable work and expenditure of resources at all levels of the organization. The new product needs everything, from a package design to warehouse space, and it all has to come out of company resources. To put forth even a modest effort to overcome the typical marketing obstacles and carve out just a small share of the market for the new product will use a sizable amount of the company's resources before any profit can be realized.

When a sales manager has to budget an appropriation of $67.71 as the average cost every time he sends a salesman on a call to a customer, or when he must come up with $25,000 to buy space and design, set up, and run an exhibit booth in a trade show, it isn't long before these expenditures can exhaust the budget. The delay also has a psychological effect. Promotional material that stays around too long begins to lose its impact and effectiveness, and the enthusiasm of the company's salesmen begins to sag.

The far-reaching effects of speed are demonstrated by the Philadelphia Quartz Company in its product development, long-range planning, and organization structure, as well as in the actual selling effort. In the early 1970s, Philadelphia Quartz had the problem of launching a newly developed antislip agent into a market in which the product did not fit into the established product line. It solved the problem by reorganization, making the research and development department responsible to the marketing vice-president. (Normally, R&D is a separate department.)

The departmental shift increased the interchange between the people who develop and modify products and those who sell them. By joining the two usually separate functions, the company eliminated 6 to 8 months of the 18 to 24 it normally took to introduce a new product. As a result, the company was able to achieve a previous ten-year projection in five to six years. Considering the long time and high cost of commercializing new products, the speed with which new products are introduced is critical.

Don't Be Reckless

Thus, speed is essential to the total marketing strategy, from product development and organizational structure to the actual promotion effort. In a few cases, a hasty marketing effort proved a disaster. There are also instances of marketing efforts that turned out successful in spite of their relatively slow execution. By and large, though, speed pays off. Overlong deliberation, cumbersome committees, long chains of command from the home office to the salespeople in the field, and poor communications all cut down on the speed of action and are detriments to the success of a marketing effort. A marketing effort may lack novelty or ingenuity, but if it is delivered forcefully and at the right point in time, it may well succeed.

In certain industries, accepted practices or traditional buying patterns force speed on marketing plans. Take the women's ready-to-wear field, where seasonal changes limit the time available for a plan to succeed. Manufacturers must get their styles accepted and into the stores by the time the season begins, or they lose most of their sales opportunity. The toy and game field is another example, for many toys and games are fads that have a relatively short selling cycle. Marketers who handle these items have to move rapidly to take advantage of the demand.

However, quick action, in the sense discussed here, does not imply recklessness, nor does it condone shoddy products or misleading advertising. It still means a coordinated, well-timed, well-executed plan that moves rapidly to exploit a market opportunity. For that matter, badly made products and false claims used in an attempt to get into the market more quickly will probably end up doing more damage to the company than a slight delay would have meant. No company can afford to sacrifice its standards, or indulge in unfair trade practices simply to get the jump on a market demand before a competitor has a chance. Guidelines for determining how fast a strategy should keep a marketing effort moving and what should be sacrificed for

Part Three Strategy

that speed must take company policy and the particular market situation into account.

Guidelines for Timing
How does a company determine the rate at which a strategy should keep the marketing effort moving and then evaluate whether it is being maintained? There are a number of checks, controls, and measurements a manager can use both to set up a timetable and see whether it is being followed or has bogged down at some point. Much of the information the manager needs is available from the company records. Further information can be developed through special market research or can be purchased from outside research sources. Some of the useful types of information and approaches are discussed in this section.

Expense. Expenditure reports are usually prepared on a weekly, monthly, or quarterly basis. These reports list the actual expenditures made during the period, as well as the amount budgeted for that period. Both figures are compared in ratio form. The expenditures are usually broken down into categories and then added as a total figure, often shown as a ratio of the total sales for the same period for further comparison.

Sales. Sales reports are of many kinds. Depending on the company and its record-keeping requirements, there may be sales figures according to time period, geographic area, individual salesmen, product category, store, department within a store—and by any other form of breakdown that the company finds useful. Actual sales figures for whatever breakdown is being used are usually shown side by side with the budgeted figures of the marketing plan. Therefore, the two can be compared to determine if profits are coming in at a sufficient rate to offset expenses.

Share of the Market. Share-of-the-market figures provide a very useful measurement if the total market can be estimated fairly accurately and the share that the company logically expects to attain can be determined. At specific times during the marketing campaign, the share of the market attained to date can be compared with the goal set for that date. What is most important, however, is that the figures be large enough to be significant. The major automobile manufacturers take share-of-the-market figures very seriously and use them very profitably in planning. On the other hand, a local dress shop would be wasting time if its management were to try to collect and use data on the small share of the garment market it could claim.

Warehouse and Retail Feedback. Warehouse and retail feedback of product movements produces valuable information about sales and distribution problems. If the product flow stalls at any point—factory, warehouse, wholesaler, or distributor—the feedback will show where the mixup is taking place. An interruption in the steady flow of the product could give competitors the time and opportunity to capture additional sales for their products. Product flow should be not only smooth, but speedy. The speed with which the distribution from manufacturer to the ultimate customer is completed is an important measure of the success of the marketing effort.

Product Acceptance. It is not enough to know that the stock has left the factory, or that it has left the warehouse. It can sometimes even be misleading to know that it is moving off the marketer's shelves into the hands of the customers. That product could still be sitting in the customer's storage room or closet, unused or discarded. The marketer wants to know the degree of customer acceptance of a product, but determining that degree of acceptance is not always easy.

A fair picture of the degree of product acceptance can be drawn by studying customer and dealer reorders. If the product is the type that should be used up quickly, such as soap powder or cereal, then reorders should start coming in very soon if the product has won customer acceptance. For a product that does not bring quick repeat sales, such as a lawn mower or a power drill, dealer reorders will indicate that the users have developed confidence in the product.

At the manufacturer's level, a simple telephone survey of dealers and wholesalers sometimes produces valuable feedback about the degree of customer acceptance of a product. If the product is a widely used consumer item, a more extensive survey can be made by telephone and personal interviews.

Advertising Effectiveness. Measuring the reaction to advertising can be a key element in determining the progress of a marketing campaign. Most marketing efforts hinge on the effectiveness of the production segment of the plan. People don't beat a path to the door of the company that puts out a better shampoo. A promotion effort has to tell them about why it is better.

Coupon returns, requests for product information, the degree of store traffic immediately after an ad has appeared, and customer surveys are some of the techniques used to measure advertising effectiveness. Although none produces hard and firm figures, each of these checks gives a fair indication of the degree of customer response to advertising and other forms of promotion.

The Product Life Cycle. The concept of the product life cycle is a useful construct for establishing the speed with which a marketing campaign should proceed and for measuring whether the effort is maintaining that speed. All successful products are born, mature, and die. Each product lives out its own cycle in its own length of time and differs in the length of time it spends at each phase of the cycle. But the configuration of the successful product's life takes on a classic pattern of evolution as it advances through the stages of the product life cycle.

The normal curve for a successful product is illustrated in Figure 8. Here, the fundamental concept of the product life cycle is pictured as the classic curve for which the vertical scale measures the sales of the product (also referred to as saturation or customer usage), and the horizontal scale represents the passage of time.

Figure 8 plots the life cycle stages of several successful products in the audiovisual electronics industry. The nodes along the normal curve indicate the current approximate position in the life cycle of each product. The two offshoots above and below the main curve represent deviations from the typical success story: a product failure and a product success. Portable reel-to-reel tape recorders failed dramatically before they ever achieved growth, when the less cumbersome cassette player was developed. On the other hand, 8-track tape decks had traveled along the typical curve into their phase of decline when a new market, or customer target, was identified for use in automobiles. The figure shows the 8-track auto tape deck sweeping off the normal curve into a new growth period.

The product life cycle concept is a convenient scheme for classifying products according to stages of acceptance in a particular market. It can be applied to such marketing activities as product planning, fore-

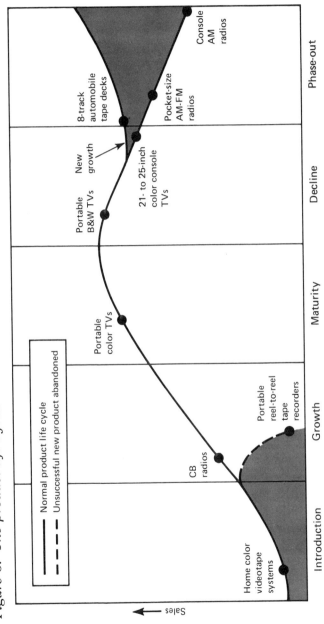

Figure 8. The product life cycle.

casting, advertising, and pricing because it provides the framework for grouping products into predictable stages. The process is not down to a science, however. It is a valuable concept that requires a certain amount of art and judgment in application.

The marketer who is about to introduce a new product should try to approximate what stage of the life cycle similar products are in. It would be foolish, for instance, to introduce a product when similar products are already in the declining phase. If similar products are in the growth phase, then the marketer can logically expect a demand for the product.

Using the product life cycle curve helps the marketer determine how quickly to make a marketing move. The marketer is able to judge how long demand will continue to grow and how long the product will probably stay in its peak phase by determining the phase that similar products are in. This tells the marketing manager over how long a period sales can be expected to grow for the product.

Once the campaign is under way, the marketing manager can see whether the product will probably be a success by the shape of its life cycle. If the typical growth pattern does not develop, the marketer might be wise to reevaluate the marketing strategy. A product that isn't going to make it should be withdrawn from the market before it hurts profits or reputation significantly. This is exactly what happened to the Ford Edsel. The new automobile simply did not develop a healthy sales pattern, one that reflected the normal life cycle of a successful product. The company finally had to abandon the effort completely, feeling that any further expenditure would be wasted.

Sometimes the marketing effort can be saved by a change in strategy. The company has to reanalyze the various elements in the marketing mix and put them together in a new way to formulate a different strategy. Once a strategy has failed, however, it should never be repeated, even if it seems to have better application to another marketing situation. This is because the strategy has already been seen by the competition, which will have had time to develop even better ways to defeat it. A discussion of product strategies is picked up in greater detail in Chapter 8.

THE INDIRECT APPROACH

A second characteristic of successful marketing strategies is the indirect approach to whatever obstacles stand in the way of the marketing plan's success. An indirect approach means taking a different path toward the desired goal. It means making a move that confuses the competition and masks the real goal it is seeking. Such an approach is used to throw the competition off balance, not only psychologically, but sometimes even physically, in terms of the way the competition has balanced its marketing mix. The indirect approach is like an end run in football. The goal is directly ahead, but the ball carrier runs off at an angle, avoiding a head-on attack on the line. He uses indirection to baffle the opposing team and gain some yardage in his fight toward the goal.

In the early 1970s L'EGGS, a division of Hanes Corporation, demonstrated the application of an indirect approach when it opened a new distribution channel for pantyhose and stockings. Prior to the market entry of L'EGGS, the main line of distribution for the "big three" companies—Burlington, Kayser-Roth, and Hanes—was through department stores and specialty shops. Hanes' strategy was to take an indirect approach, to bypass its competition and enter a new distribution channel through the supermarkets and drugstores. The company created a brand name, L'EGGS, used a unique package, and set up an attractive display rack. It also set up a network of route girls to service the racks.

Hanes went even further when it took the unorthodox approach of asking the store managers to allow 2½ sq ft of space for the display and to take the hosiery on consignment. The practice of consignment selling was generally avoided by other merchandise sellers, specifically the inexpensive foreign hosiery brands, because consignment selling does not require an investment or risk on the part of the store manager. Two years after the marketing effort was launched for L'EGGS it was selling 75 percent of the major urban markets in the United States and had achieved first place as the largest-selling brand of hosiery in supermarkets and drugstores, with sales of $35 million.

The Profits Reaped by Unorthodoxy

The indirect approach is part of that valuable planning device—unorthodoxy. Unorthodoxy defies the traditional way of doing things, the rigid thinking, the set rules, the by-the-book approach. It represents a new twist and depends on originality and surprise to achieve success. Enterprises that have violated the traditional ways of making marketing moves and used the indirect approach have proved how well such methods work.

According to the old rule, to sell phonograph records during the early 1950s a marketer had to have a store in which the customer could listen to each record before making a purchase. Not true, said some unorthodox companies led by Columbia Record Club—which proceeded to prove its point by selling millions of records through the mail.

For a long time, pharmaceutical houses operated under the assumption that proprietary drugs could be sold successfully only through drugstores. A few companies challenged that idea, and now a number of drugs are sold successfully in all kinds of outlets, in supermarkets and through the mail. One market that had seemed almost impenetrable was the soft drink market, dominated by such giants as Coca-Cola, Canada Dry, and Pepsi Cola. Schweppes managed to find a weak spot in one segment, used a lively promotional campaign as a wedge, and won itself a narrow, but profitable, share of the business.

Detroit was sure that small cars were just a fad until the energy crisis hit. Volkswagen had disagreed all along and kept its prices competitively low by ignoring the tradition that car styling had to change every year. Sony entered the U.S. television market not by tackling such giants as RCA, Westinghouse, and General Electric head-on, but by using the indirect approach and carving out a profitable piece of the miniature TV business.

What's Wrong with a Direct Approach?
Strategy based on a direct approach, on doing the expected, often makes a company particularly vulnerable to the competition. The direct approach usually means using standard tactics and well-known ideas and copying current fads in advertising themes or product design. Doing the predictable often results in a marketing effort that looks dull and too me-too-ish to prospective customers. It gives the competition an

opportunity to woo them with a more interesting approach.

Managers who insist on using direct methods are usually overly concerned with making mistakes that might be held against them. They don't want to take chances, so they use the traditional tactics, the obvious strategy, and, of course, their marketing moves are obvious to the competition. Direct methods simply don't pay off.

The unexpected approach will baffle the competition. They may react by making mistakes which translate into opportunities for the challenging company. In most cases, an indirect approach will throw the competition off course only temporarily. But even a brief upset may be long enough to give the company the chance it needs to prepare for the next move. It seldom lasts long enough to eliminate the competition altogether from the market, but it is highly unlikely that a marketing effort would be directed toward such a goal. Few companies have the resources that such an effort would require, and any marketing company that even seems to be attempting to monopolize a market soon finds itself in the federal courts.

UNBALANCING THE COMPETITION

A third characteristic of successful strategies in marketing plans is any move that puts the competition off balance and thus weakens the reactions the competition can muster against the developing marketing campaign. Like a multiple offense in football, its goal is to cut down the effectiveness of the competition's response by spreading it out as thinly as possible, by diverting and confusing it.

Part Three   Strategy 181

An effort to put the competition off balance must be done at the lowest possible expenditure of company resources, as with all other moves in strategy. A manager must never forget that the extent to which any marketing plan is successful is measured in the financial terms of net profit or return on invested capital. The move to "unbalance" the competition may be the most important piece of strategy in the entire marketing plan. This is the move that forces the competition to react defensively and *change*. Remember that paradox? Competition is usually listed as an uncontrollable factor, and yet to change it is the goal of most marketing efforts.

Types of Moves
The moves that cause the competition to reassess its situation can involve physical or psychological action, often a combination of both. If the competition is successfully holding onto an important share of a market or making a strong bid for a bigger piece, it must be because that competition has found the right marketing mix. Cause an imbalance in that marketing mix in any way whatsoever, and it is likely that the competition's reactions and market position will be weakened.

The Physical Maneuver. A physical move to weaken the competition's position usually takes into consideration the market conditions, the marketing mix of both companies, and some careful timing. Based on the condition of the market, it usually involves shifting some physical resources at a time and in a way calculated to force the competition into making a quick countermove.

For instance, a company might have a marketing objective of increasing its share of a particular two-state market area from 12 percent of the total sales made to 18 percent. Standing in the way of that objective, however, is a strong competitor that seems perfectly capable of holding onto its share of the market. The company's strategy, therefore, might include an attempt to change the market situation by focusing a maximum sales promotion campaign on those two states, using increased advertising and shifting additional salesmen from other territories. In addition, the company might set up improved facilities in the area for fast product distribution to assure dealers of plentiful supplies and quick delivery of reorders.

Even though these moves would be obvious in their intent, they would still be able to change the market situation. The competition would then be forced to change its marketing mix to meet the challenge. For instance, a competitive marketer might also step up the advertising campaign. It would represent a switch from what had been such a successful marketing mix for the competition, a change in the balance of product, price, promotion, and distribution.

The Psychological Maneuver. Psychology plays its part, too. Often a manager will make a marketing move that may not directly challenge the competition, but is intended to make the competition uneasy and unsure. A series of teaser ads saying, "Wait until you see what we've got for you in September," would start the competition wondering whether the innovation is going to be a new product, an improvement or change in a standard product, a change in price, or. . . . Again, the move, although only psychological, has its unsettling, disconcerting effect.

With its brand of L'EGGS hosiery, Hanes circumvented its major competitors and gained a major market share through supermarkets and drugstores; this had an unbalancing psychological effect on the "big three" company managers. In addition, it would be reasonable to assume that the marketing managers of the low-priced, foreign brands of hosiery already in supermarkets felt the unbalancing effect of the fast-moving, innovative, and indirect approach of the L'EGGS strategy.

The Decision
The deciding factor in knowing what move will most disconcert the competition is a determination of just where and how the competition is vulnerable, what is needed to take advantage of that vulnerability, and whether the results are likely to be worth the cost. The purpose of the strategy is to get the competition to react in some way other than its traditional strength. The hope is that the competition will spread out its resources and disrupt its marketing mix, thus changing its marketing strategy. Even if the strategy of unbalancing the competition does not in itself achieve the final objective of the marketing effort, what it does is to confuse the competition and prepare the way for future attainment of that objective.

A Case History
Here is an example of how one company might try to unbalance a competitor to reduce its growing influence in a market the company was determined to keep. The Pemco Company was a manufacturer and wholesaler of a line of washroom products for industrial plants and office buildings. Sales records suddenly began to show that Pemco was losing part of its business to a competitor, the Shenker Corporation. Pemco management studied the market situation, collecting as much information as possible about the marketing

mixes of the two companies. On the basis of this research, Pemco came up with a forecast of what it thought Shenker's strategy would be for the coming year. Management decided that Shenker probably would:

- Cut prices on existing products that were most strongly competitive with Pemco products and try to get Pemco to do the same.
- Add new distributors by offering them larger discounts than those currently being given, with a particular emphasis on trying to lure away Pemco distributors.
- Develop and put on the market a line of lower-cost products in order to gain more flexibility in competing on a price basis.

The way Pemco saw it, Shenker would emphasize price, and knew that it could not match the Shenker price tactic. Therefore, Pemco decided to try to unbalance Shenker's marketing mix by using a carefully worked-out strategy: Pemco decided to:

- Avoid getting into a price war with Shenker, even though such a policy might mean losing some sales volume. At least it would avoid adding to the unit cost of manufacturing the products and, when possible, even cut those costs.
- Continue to give distributors the same discounts, even though the policy might cause some distributors to decide to shift to the Shenker line.
- Concentrate resources on strengthening the distribution program by upgrading and enlarging the sales force, working on improved distributor relations, and streamlining the physical distribution and warehouse network for quicker service.

Pemco decided to stay out of the price war that Shenker wanted to develop. Instead, Pemco used its resources to strengthen the distribution of its marketing mix, in the hope that Shenker's plans would be thrown off balance, perhaps even necessitating a change in strategy. Pemco kept its share of the market by using physical resources and careful psychology in a way that Shenker did not expect.

The Marketing Center of Gravity
When attempting to unbalance the competition, the important element to look for and isolate, in both the competition's marketing mix and in the company's own marketing mix, is what is called the marketing center of gravity. The marketing center of gravity, or MCG, is the pivot in any company's marketing mix, the strength on which everything else depends.

The MCG can be used in any area of the marketing mix. It can be the product itself or the advertising that the company uses. It can be the personal selling techniques developed by the company or the type of distribution outlets the company has. These are just a few examples. Every company's marketing mix, however, has its own particular strong point, its own focus of strength, its own MCG on which the company's success depends. To locate and study a company's MCG, a manager needs:

1. A *market information system* that supplies regular, continuing, up-to-date data about the market and the companies competing in the market. The information supplied should be complete enough to allow a manager to anticipate problems, identify opportunities, and prepare marketing strategy;

2. An *organizational structure* with rapid lines of communication between the men in the field and the

decision makers in the home office, a communications network that cuts through all the layers of management to get the needed information to the right people as quickly as possible.

Finding out exactly what a competitor's MCG is helps a company shape the plans that can unbalance the competitor's position. What the company wants to do is to reduce the effectiveness of the pivotal point, that MCG, so that all the other parts of the marketing effort will be less effective as well. The MCG can be made less effective by matching marketing mixes or emphasizing different mixes, depending on which tactic a manager thinks will be most successful without increasing costs. In the Pemco example, it was determined that Shenker's MCG was its pricing policy. Pemco decided not to match, but to mix by improving its own distribution system to a point at which Shenker would have to take valuable time to divert its resources away from the price sector and use them to strengthen its distribution sector.

CONCENTRATION

The successful marketing strategy characteristics discussed so far are speed, indirect approach, and unbalancing of competition. The next characteristic is also the next logical move in strategy—a concentration of a company's strongest resources against segments of markets and at the weakest point in the competition's marketing mix.

A concentration of resources means focusing money, manpower, and materials of a company at segments of the market where there is an identified need for a product or service. It also means concentrating resources where there are one or more weaknesses in the competitor's marketing capabilities. For example, there could be some deficiency in the product, or inability to provide adequate distribution and service, or ineffective promotion to move prospects to the point of purchase, or inability to price the product or service to gain a major share of the market.

The mark of a successful marketing manager is the ability to judge his or her own company's marketing resources and then be able to compare strengths and weaknesses against those of the competitor. It is to determine whether it is wise to enter a market as measured against the accepted standards of sales, profits, and return on invested capital. Sometimes a weakness in the competition's marketing mix is obvious at the beginning of the marketing effort, and the manager's preliminary moves are made simply to increase this weakness.

The marketing manager at L'EGGS, for example, saw that even though the competing foreign hosiery brands were low priced, they were also of low quality, the brand names were not nationally advertised, the competitors' packaging and display lacked excitement and eye appeal, the store manager had to make an up-front financial investment in inventory, and above all, the other leaders in the hosiery field, Kayser-Roth and Burlington Industries, did not fill the void that existed at the time of Hanes' entry into supermarkets and drugstores. Thus, two situations existed for a concentrated marketing effort. First, there was an identified market need for quality, brand-name hosiery conveniently available through supermarkets and drugstores; second, there were weaknesses in the competitors' marketing efforts.

Unless a company has a monopoly in the production of a product or in its distribution, both of which are unlikely in the normal marketplace, it cannot fill all the distribution channels or reach all the potential customers. Therefore, it must have weaknesses somewhere, which means opportunities always exist. Weaknesses can be found almost anywhere in the marketing mix, and opportunities can arise almost anywhere in the marketplace. It could be a poorly trained and ineptly managed selling staff, a lack of proper quality checking on the production line, or a case of ineffectual advertising; or perhaps the company has overexhausted itself financially. Any one of a combination of these conditions would constitute a major weakness that could be exploited by the right kind of marketing strategy.

A few years ago, Philip Morris, Inc. sensed an opportunity for concentration of its selling and distribution. To substantiate the hunch, it launched a massive computerized study covering every retail outlet in the country. The study ranged from giant supermarket chains down to the corner mom-and-pop store. With this huge mass of data, Philip Morris, Inc. isolated its high-volume outlets that required four calls or so per month. Then came the medium-volume outlets that might be worth one or two calls per month. Finally, there were the low-volume retailers, who might not be called on at all, except once or twice a year during special sales drives. At this point, the company examined the location of the outlets and was able to concentrate its workforce at points of high sales and profit opportunity. Prior to the study, there may have been three men in North Dakota and three in San Francisco, where the best concentration of selling power would have required only one in North Dakota, four in San Francisco, and perhaps another in Los Angeles.

Furthermore, the application of concentration was part of the strategy. By examining the number and location of warehouses, Philip Morris was able to consolidate and concentrate a warehouse network that achieved warehouse delivery to sales outlets on an overnight basis. Like all other characteristics of good strategy, the strategy of concentrating strength against opportunity is more of an intellectual device than it is a physical encounter. It is not only a matter of which company has the better salesmen, the bigger ad budget, or the larger manufacturing capabilities, but it is also a question of how and when a company manipulates these resources in the marketing effort.

Shape of the Market
If strengths are recorded as peaks on a chart, and weaknesses are recorded as valleys, then a company's marketing mix takes on a definite shape. The prudent manager bases a marketing strategy on what he or she finds to be the "shape" of the competition and on how the shape of the markets is plotted. The various information systems, whether from informal sales reports from men in the field or through formal market research, give the manager the detailed information needed to determine the shape of the competition's marketing mix, as well as keeping an up-to-date picture of any changes in the market. Philip Morris' study helped the company determine the shape of the marketplace and how to allocate resources to conform to that shape.

Only when a company has manipulated its competition can it know the best point to concentrate on—what point would be the least likely to withstand a concentration of effort against it. However, managers want to keep the competition from doing exactly the same thing to their companies. They want to disguise their marketing plans and strategy. Managers try to keep the competition from tracing the shape of their marketing mix and from working out some means of exploiting its weakness. When preparing for a concentrated effort, good strategy often includes preliminary moves.

For example, a manufacturer prepared for a concentrated effort in a particular geographic area might introduce a new product in another area simply to mislead the opposition. The distraction might well draw some of the competition's strength away from the market area, and increase the chance of success for a concentrated effort. A small company with limited financial, production, and marketing resources might wait for the industry leaders to bring out a new product. If it becomes a success, the small company can then turn out a version of it and hunt for a piece of the market. To get that chunk will take clever strategy—but the small company will have saved its resources, having spent little, if any, in the development and testing of the product idea, and will have learned a good deal about the marketing mix of the successful larger company.

Still another version of this type of strategy is for a small company to indicate, through deliberate information leaks and promotional activity, that it intends to market a new product. This might lead a major company to react by trying to get a similar product onto the market first. This gives the smaller company the opportunity to check the profitability and consumer acceptance of such a product and to study the marketing mix of the larger company.

The success of the strategy of concentration of resources usually depends on a definite plan for thinning out or weakening the competition's strength before the effort takes place as well as examining the shape of the market. Shaping the competition shows a company exactly where the competition is weakest and where the opportunities are the greatest.

Cumulative Concentration
Although the concentration of resources against the competition and market segments is one of the key elements of a successful strategy, it also can be a dangerous technique unless it is thoroughly understood and carefully used. Simply pouring out all the available money, manpower, and materials at one point is not going to succeed unless the competition is unable to defend itself at that point. The wise manager commits his or her company to such a course of action and uses its strength only when the situation assures a good chance of success.

Sometimes success requires a cumulative action rather than a single effort. The concentration of a marketing effort does not necessarily mean taking all the resources and applying them with one tremendous thud against one point of the competition's marketing mix. It does not mean, for instance, that the entire budget should be sunk into one promotional effort, nor that one product should be relied on to carry the entire weight of achieving the marketing objectives.

Instead, a concentrated effort should be launched. A single ad, a single salesman, or a single product improvement is not enough to make the marketing effort succeed. The effort usually combines the company's strongest assets directed in a series of cumulative programs. Strength is multiplied by a cumulative effort. For instance, a single ad in a trade magazine may cause a purchasing manager to remember the product for a few days. An ad followed up by a sales letter may keep the product in the purchasing manager's mind for a few weeks. An ad followed up by a sales letter and then by a call by a salesman will probably fix the product firmly in the mind of the purchasing manager, and may result in a sale.

ALTERNATE OBJECTIVES

The final characteristic found in successful strategies is the existence of alternate objectives. These should be both sequential and alternative. In other words, the strategy should involve a series of objectives to be attained, one after another, and there should be alternative objectives at each step.

Alternate objectives give the plan flexibility and allow a company to react to changes without losing any headway toward the marketing effort's ultimate goal. If a plan has but a single objective, it may prove impossible to reach, but then the entire marketing effort has to be written off as worthless. If it has a series of alternate objectives, the company has a good chance of obtaining at least one objective and using that success as a stepping-stone toward the next step.

Alternate objectives also help confuse the competition. If the company has only one objective, the competition probably will learn what it is and use all its strength to defeat the marketing effort. If, on the other hand, the company has a series of alternative objectives, the competition won't know which area to defend. No matter how competing defenses are moved, the company will always be able to switch its marketing effort to an alternative objective.

Put in practical terms, it is the case of the hardware store owner who wanted to get some of the construction hardware business but wasn't sure of what kind of supplies to handle. What the management wanted was an increase in sales volume through the development of a specialty product line. After studying the market, the strategy that was arrived at involved a choice of objectives, which could be reached either by adding a comprehensive line of plumbing hardware or by adding a full line of interior fixtures.

First, the store brought in and promoted a relatively modest stock of both plumbing hardware and interior fixtures. Volume gradually increased, and it becane clear that most of the increase was in the plumbing hardware. A study of the total sales figures and the market situation showed that most of the interior fixtures business was still going to a firmly entrenched local company, but that none of the other companies offering plumbing hardware had built up any particularly strong hold on their customers. The hardware store then dropped its line of interior fixtures, concentrated on plumbing hardware, and achieved its goal of developing a profitable specialty.

TWO ZONES OF ACTIVITY

A company's marketing plan strategy involves winning more business, usually by combating competition in a particular market. It is equally important, however, that a company's strategy take into consideration the wants and the needs of the customer in terms of both today's and tomorrow's environment. As the social, political, and economic environment changes, so does the customer. In the end, satisfying the customer is the reason the marketer stays in business. Here, then, are the two zones of strategy activity that the manager must balance: the competition and the customer.

The Competition

Making a marketing plan and formulating its strategy to defend or develop a market requires decisiveness, resoluteness, and firmness. The keys to success in formulating a strategy can be summed up as follows:

First, adjust the objectives to the available resources. In determining a marketing objective, calculate what is possible to achieve. Do not take on more than the company logically can handle. Work with the full realization that, as the marketing effort develops, unexpected situations also will develop that take more of the resources than originally planned. Therefore, don't figure too close to the limits of available manpower, money, and materials in any marketing effort, and don't plan a marketing effort that involves an unrealistic and probably unattainable objective.

Second, keep the objectives in mind at all times. Adapt the plan to circumstances, realizing that there is more than one way to achieve each objective. The plan should be flexible—and every action should take the company a little closer to the final objective.

Third, choose a course that the competition does not expect. Try to determine what course would both surprise the opposition and catch them off guard as well as take the company closer to the final objective, for that course will be the one that the competition will find hardest to defeat.

Fourth, take the line of least resistance. Don't batter against the competition's strongest point. Concentrate instead on a weakness in their marketing mix, just as long as the action contributes to the movement toward the ultimate market objectives and the company can use its strength efficiently.

Fifth, develop a plan that includes alternate objectives. By putting a competitor in the position of not being sure what kind of marketing effort to expect, the company has the chance of gaining at least one objective and may be able to use that success to reach for the next. But don't confuse the difference between the value of establishing a single, coordinated marketing plan, which is a must, with picking only a single objective, which is dangerous.

Sixth, don't repeat a plan or strategy in its same form if it fails. Simply adding more resources to an unsuccessful strategy will not succeed. Not only will the competition know what form the marketing effort will take, but the time it takes to reinforce the effort will give the competition time to strengthen defences. And finally, keep these tips in mind:

- The more strength a manager wastes when entering a market, the greater the risk that the competition will be able to defeat him. Even if a manager eventually succeeds in winning the market, he will have eaten up a large part of his possible profits in the process.
- Although the tools of the marketing mix are physical (product, price, promotion, and distribution), putting them together into a marketing mix is a mental process. The better the manager's strategy, the more likely he is to gain the objectives of the marketing plan and the less that victory is apt to cost him.
- The closer to the line between what is ethical and what is nonethical a manager's marketing practices are, the more bitter he will make the competition, with the natural result of hardening the very resistance he is trying to overcome.
- If the contest for a share of the market is going to be too costly, the manager should abandon it or should rework his strategy to bring the ratio of cost and profit into line.

The essential sequence in establishing a marketing strategy for a marketing effort is to first create the opportunity, and then exploit it. A company cannot enter a market or hold onto its share of a market unless it first creates the opportunity to do so. It cannot succeed in its objectives unless it then exploits that opportunity. In other words, success doesn't just happen, it is planned.

One final comment about the competition: Remember, the objective of a marketing manager sometimes has to take into consideration the need to learn to live with the competition and protect the profits that the market is providing for every company in it. Excessive discounting and price wars don't really change the market for the customer, but they do eliminate the chance for an adequate profit margin for all companies involved.

The Customer's Environment

Marketing plan strategy should always relate the product to the customer and the environment in which the customer lives. Marketing objectives go far beyond profits and sales. They must involve more than merely satisfying today's customer and doing it at a profit. The goals must also take into consideration what tomorrow's customer is going to want and how tomorrow's environment will affect the decisions of that customer.

A company has both an involvement in and a commitment to the environment in which its customers live, for environmental changes shape market changes. First, environmental changes can cause problems in what have been established and profitable markets. For instance, companies that make cigarettes are facing increasing promotion restrictions and increasing social disapproval because the product they offer has been proved to be detrimental to people's health. Companies that manufacture firearms for civilian use are facing a demand for federal and state legislation that would not only sharply restrict their marketing activities, but might put many of them out of business. Even toy companies have their problems, for various groups would like to see the production of toy war games and machines curbed. Although technically these companies are practicing the marketing concept by producing and selling a product that satisfies a specific group of customers, a change in the environment is putting steadily increasing pressure on them that may eventually change the entire market picture.

At the same time, environmental changes are creating new opportunities. For instance, some toy companies have put out lines of multiethnic toys, such as dolls with black features and Afro hair styles, and have discovered that these lines have not only satisfied the social responsibility of the companies, but have brought in increased sales as well. Other companies are developing and marketing products and services to handle some of the major concerns of the day—new ways to treat illness, to build low-rent housing, and to cut down air pollution and water pollution. As one marketing expert has said, "One of the next marketing frontiers may well be related to markets that extend beyond mere profit considerations to intrinsic values, to markets based on social concern, markets of the mind, markets concerned with the development of people to the fullest extent of their capabilities." Good strategy will be the essential catalyst for plans concerning these new markets.

Chapter 8
Activating the Marketing Strategy

Marketers must continually take the initiative to obtain and analyze information and use it to develop plans and strategies that make the best use of the company's resources. Marketing managers must make their decisions and perform their functions in response to the dynamics of a volatile and increasingly competitive marketplace. Even though typical marketing decisions are made in an atmosphere that often demands an on-the-spot, on-target response, the planning process is essential. Continual reappraisals of assessments and assumptions are made as new situations arise. Plans and planning represent the only shelter available for the marketer to work in while making decisions.

PLANNING THE MARKETING MIX

As close as the words are, there is a big difference between a "plan" and "planning." A plan is a static entity, a collection of facts and figures. It lays out the details about what has been accomplished and what is still to be done, and it presents goals against which to measure progress and results. Planning, however, is action—creating a marketing plan, choosing the plan's strategy, in fact the total activity involved in implementing the plan.

Planning is an ongoing process in the operation of any company. As such, it is much more important to the company than is any single plan. One plan is just one blueprint, one commitment, but planning is the action that begins with the original decision to make a marketing move that continues right through the final evaluation of the results achieved. Managerial planning as discussed in the previous chapters covers the ways in which a marketing plan and its strategy are developed: how data is collected and analyzed, what the elements of a marketing plan are, and how strategy is worked out to weld the plan into a strong, action-ready, competitive weapon. Now comes the final phase of planning—putting the plan into action and implementing the marketing mix.

While a plan is being put into action, planning activity continues at all levels of management. Top management continues to do its share of planning: reviewing and updating long-range objectives and strategies in accordance with changing market conditions and the results the marketing efforts have achieved for the company. Middle management contributes its part of the planning effort; making the day-to-day, week-to-week, month-to-month decisions needed to keep the marketing effort moving smoothly, speedily, and in the right direction.

The action in implementing a marketing plan involves one, several, or all four ingredients of the marketing mix—product, price, promotion, and distribution. The action, for instance, may be concentrated on introducing a new product design, developing a new pricing policy, launching a new promotional campaign, realigning the means of distribution, or perhaps a combination of all four. Whatever the action is, it takes planning to carry it out effectively. Here, then, are some of the ways in which planning puts a plan into action both through the marketing mix—and the makeup of the market itself.

PRODUCT STRATEGIES

A product may be defined as "a collection of benefits housed within a physical framework." The key word in that definition is "benefits." Unless a product has benefits in the eyes of the customer, it will not satisfy a customer need or want. Yet how well a company can satisfy customer needs and wants, and do it at a reasonable profit, will determine how successful that company will be.

If a company decides to concentrate its marketing effort on a product, the action may involve developing sales for a new product or strengthening sales of a product already on the market. Sometimes the action is planned to accomplish both jobs. For example, a new desk calculator can be promoted by its manufacturer, not only for its own usefulness, but as a valuable addition to the line of office equipment that the company also offers.

New Products
New products are the lifeblood of business growth and prosperity. Eliminate the process of coming up with

new products, be content to stay with the same old product lines, and the marketer might just as well get the "going out of business" sign ready. Yet the process of coming up with a need-satisfying and want-satisfying (and hopefully customer-satisfying and profitable) new product is becoming increasingly difficult every year. Seven to nine out of every ten new products introduced over a given year do not become commercial successes. That very sobering statistic means that careful planning must accompany every step of the marketing effort for a new product.

Good planning uses its organizational structure to the best advantage. The process must involve thinking in terms of new products, following a careful procedure of generating and testing product ideas, and knowing when and how to launch a new product to give it the best chance of success.

Administrative Organization. Unfortunately, too many companies treat new product development as a haphazard affair, reacting to competitive efforts instead of careful planning: Company A brings out something that seems new and profitable, so company B works out a variation of that product to get some of the market. Planning for new products should be an integral part of a company's organization. A formal system should be set up that encourages the company to initiate action rather than simply react to what other companies produce. A widely used system of organizing for new product planning is the product-planning committee. This committee should have a formal status within the company organization. It should meet regularly and be expected to show results.

If the company is a manufacturer, the committee members might be from the production, sales, and finance divisions. In a retailing company, the broad divisions of merchandise carried by the company might be represented. For companies in which technology is important, it is vital to have a research and development expert on the committee, someone who is not a laboratory worker, but who is in direct contact with the marketplace and with the various sales divisions. In those companies that handle large quantities of consumer goods, a product manager or brand manager can be useful in pointing out where a new product might fit into the company's existing array.

The size of the committee is not important. It is a function of the size of the company. Whether it is a team of two or a group of eight, the committee has to have the full approval and active guidance and participation of the company's top management. Without the interest and backing of top management the committee will be useless.

Developing New Product Ideas. Once the organizational structure has been worked out, the committee must set up its working procedures. The development of a new product begins with consideration of the ultimate customer. The committee's goal is to pinpoint a customer need or want, and to seek out or develop a product that will satisfy it. The exact sequence of steps involved in such a system varies from company to company, but the steps most companies use are as follows.

First, a steady flow of new product ideas is encouraged. This means opening up channels of communication with salesmen, service personnel, and other company employees. New ideas emerge through talking with executives at other distribution levels in the industry, meeting with the key people in trade associations, and keeping an eye out for what the competition is doing. The idea is to encourage people to talk about what they think is needed or wanted in the marketplace. Once people know there is an interested ear, they tend to report anything they think may be of value.

Next, the committee must arrange a means of doing an immediate preliminary screening of all the ideas that are generated. Obviously, only a few of those ideas are likely to hold a real possibility. All the rest should be discarded or filed for future rescreening. Speed is essential here, and although it is always possible that too quick a screening can cause a possible winner to be lost, it is still better to discard even the "doubtfuls" and "possibles" and use the valuable time saved to examine what seem to be good ideas.

Now comes a crucial step. The ideas that have passed the rough screening must be studied carefully. Time should not be wasted in this study, but a sufficient amount should be taken to guard against carelessness and miscalculations. Here are some of the questions that must be considered:

What are the size and location of the potential market for the product? Is there really adequate demand to warrant its consideration? The size and scope of the market must be considered before anything else, for if an adequate potential market doesn't exist, it would be a waste of time to consider the problems of production, financing, and marketing.

Does the product fit into the company's present production facilities or product assortment? For instance, if a company makes ashtrays and sells them directly to retail outlets, adding a line of imported cigarette lighters might complement the assortment the salesmen could offer customers, but it would do nothing to help fill in slack periods on the production line. On the other hand, developing other kinds of glass and pottery gift items might increase the production and marketing capabilities.

Should the product be manufactured by the company itself, or should it be purchased from another company for resale? The costs and possible profit margins involved in the alternatives should be compared, taking into consideration the distribution timing as well. While this would seem a question of concern only to manufacturers, an increasing number of retailers now own or lease production facilities.

How much financing would be needed? A careful analysis of return on investment should be made, for this is the basic factor in determining the possible profitability of the product.

Are there any obstacles to securing a patent, if one is needed? Can the product be designed to meet the applicable packaging, labeling, and quality standards?

Would the product help the company achieve its long-range objectives, in terms of product assortment, company image, and return on investment?

If the particular product idea still seems like a good one, it should be developed to the point where a limited quantity or a prototype can be tested on the market. If the results of the completed test prove discouraging, the company can still abandon the idea and charge the cost of investigating the product idea as a general expense of product planning. If the tests are successful, however, the company should put its resources firmly behind making the new product a profitable part of its line. Because this step is so important, testing should be done in areas and with customers truly representative of the particular market as a whole. The testing should include not only the product, it should have its projected packaging, pricing, promotional theme, and method of distribution.

Launching a New Product. Launching a new product in a market means that the marketing effort will involve all of the factors in the market environment. Some, remember, are controllable, and some are not. For example, a company has control over its own marketing mix: the product itself, its pricing, its promotion, and its distribution. It cannot, however, control what customer attitudes and tastes are going to be, nor what legal restrictions may be adopted, nor what the competition will do.

The biggest obstacle in any market is the competition. A competing manager has exactly the same choices open to him in his planning as does the manager of the company with the new product. The competing manager will try to unbalance the marketing mix surrounding the new product's introduction in order to keep that product from reaching whatever objective the company has set for it. The competing manager will try to put just as much effort into trying to keep or lure customers away from that new product as the company will put into trying to get customers to buy the product.

The winner will be the company with the better marketing strategy. The company introducing the new product wants a strategy that will protect and assist the new product. It wants a strategy that will help the product quickly secure a good market position. It wants a strategy that will make it seem clear to competitors and customers alike that the new product is a formidable entry in the market, that it has uniqueness, benefits, and value.

There are several types of basic strategies used to

launch a new product. The important similarity among them is that each ties the new product, in one way or another, to the existing lines offered by the company. By doing this, the product gives added strength to the assortment, and vice versa. In three of the most basic strategies, the marketing effort is planned to underscore how the new product:

1. *Enhances the present product lines offered by the company.* The new product may expand the company's offerings, by being an addition or complement to products already in the company's assortment. A bank adding a new service to its already existing line of services is using an expansion strategy. So is the supermarket that adds racks of small packaged toys to the lines it already carries.

The new product may also condense or simplify the company's lines. It may be a new product that replaces several products in a line, or it may be a new brand or category that replaces several brands or categories carried by the company. An electric blender that can chop ice, shred meat, and keep the mix warm would be such a product.

2. *Is an improvement of an existing product.* One of the most successful strategies is to introduce a new product as an improved version of what has been available. Usually the improvement is in the product itself: it lasts longer, performs better, does more. Sometimes, however, the improvement is in the packaging, or the pricing, or the promotion, or some other part of the marketing mix, particularly if the product is for a market in which there is a great deal of similarity among competing products.

For example, a bank might establish new banking hours, staying open one night a week or remaining open on Saturdays to attract people who are at work during regular banking hours. Such a move would provide an added banking convenience, a definite plus for a very profitable promotional campaign.

While an improvement on an existing product or service doesn't allow the company the claims of uniqueness that can be made about something completely new, it does link the past popularity of an old standby with the value of trying the new product. In a way, it helps extend the life cycle of a product by changing its marketing mix just enough to rekindle new interest in it.

3. *Helps the company project a new price image.* A new price image involves trading up or trading down. "Trading up" means increasing the quality and average price of the product lines carried by the company, and "trading down" means decreasing both the quality and the average price of those lines. The most familiar examples of this strategy are among retailers. Korvette's, once a pipe-rack discount store, still emphasizes its bargains but also carries higher-priced merchandise and has traded its image up to the point where it has chandeliers and carpeting in its Fifth Avenue branch. Tiffany, on the other hand, once known only for its higher-priced jewelry, now advertises that it also carries good buys at moderate prices.

Existing Products
With all the excitement generated by new products, never forget that many existing products on the market may still be profitable best-sellers. Jell-O has been around since the 1930s as has Scotch tape. Nylon was introduced in the 1940s. Although there have been important new entries in the dessert, transparent tape, and synthetic fabrics fields, these old faithfuls continue to produce substantial sales and profits for their

marketers. However, it isn't just that a few products happened to continue to remain good sellers. The long life of these products is the result of carefully planned action, action involving specific moves made at specific stages in the products' life cycles.

Life Cycle of the Product. As discussed in Chapter 7, the typical life cycle of a product follows the predictable pattern of introduction, growth, maturity, decline, and phase-out, shown in Figure 8. The order in which these stages occurs doesn't change, but the length of time a product takes to pass through each stage can be modified by the astute marketer. What marketers have learned to do is to extend the length of the cycle. The Yo-Yo and the Frisbee are examples.

When the product is in the average introduction stage, there is a slow but gradual rise in the number of sales made. How sharply the sales line rises depends on how quickly the product becomes known and accepted by customers. Price may have a lot to do with that acceptance, for customers will try a lower-priced product more readily than a higher-priced product. As soon as a new product begins to build sales, the competition watches it closely, trying to determine whether it would be profitable to get into the market.

If the product is slated for success, it will show a sharp rise in number of sales at the growth stage. This is normally the sharpest increase in sales found during the entire life cycle. At this point, the competition begins to pay very close attention indeed, ready to try to get a piece of the business if the market looks like a good one. In addition, the first pressure from price-cutters and discounters begins to be felt, as the low-end marketers try to make profits on cheap copies. During the maturity phase of the product life cycle, the sales line levels off but stays relatively firm. Quality product variations and refinements put out by the competition make bids for the market, and price-cutting and discounting become more intense.

When the sales curve begins its downward turn, the product is in its stage of decline. Some companies pull out of the market completely, satisfied with whatever profits they have made. Other companies, even the higher-quality, prestige enterprises, begin to cut prices on what stock they have left, in an effort to get rid of the inventory, or their competition, in any way possible.

A product is usually phased out when it is no longer available in any outlet at any price, usually because there simply is not enough demand left to make stocking the product worthwhile.

Extending the Life of a Product. During the maturity stage of a product's sales cycle, when the decline stage appears to be just ahead, a marketer has the option of getting out of the market or trying to extend the life of the product. Most companies follow the first course, turning their attention and resources to the development of a different product, with a new life cycle. Some companies, however, develop a careful strategy that extends the maturity phase and delays the decline. This can produce a dramatic increase in the length of the product's life cycle and in the profits earned for the company. Here are four basic ways a company may be able to extend a product's life and thereby increase its return on investment.

1. Promote more frequent usage among current users. This means increasing the amount each present customer buys. In the case of nylon stockings, Du Pont once was troubled by a trend toward bareleggedness, so it developed new lengths and better proportions for stockings to make them more readily acceptable to women. Jell-O went from "six delicious flavors" to well over a dozen flavors, encouraging housewives to buy and serve in more variety. The 3M Company introduced tape dispensers as a regular part of the transparent tape purchase, making the tape easier to use and causing people to use it up more rapidly.

2. *Find new users for the product.* Du Pont tried to encourage the use of stockings by more teenagers and subteens, groups that usually stick to socks. Jell-O went after the weight-watching market. The 3M Company brought out both a lower-priced line to reach a new segment of the consumer market and a line of industrial tapes to tap that market.

3. *Find more uses for the product.* Du Pont added textures and colors to the assortment available in stockings, giving them sports appeal as well as dress appeal and encouraging the idea that it is fashionable to coordinate stocking colors with clothes. Jell-O showed housewives how to make salads as well as desserts out of its products. 3M brought out colored, patterned, waterproofed, and write-on tapes, which increased the versatility of their usefulness.

4. *Find new uses for the product's basic material.* Du Pont adapted nylon for use in wigs, tires, rugs, and a host of other consumer and industrial goods. Jell-O promoted its product's value as a health product, a fingernail toughener. The 3M Company turned out double-coated tapes, which competed with liquid adhesives, and reflecting color tapes, which competed with paint.

When sales of their products appeared to have passed through the maturity phase and to be on the verge of a decline, all three companies did something about it. The action was the result of a careful planning effort that had been going on since the introduction of each product. This planning enabled the companies to gauge when a change in marketing strategy should take place and to decide what the change should involve. The results: products that remain competitive, profitable, alive, and well in the marketplace.

PRICE STRATEGIES

The second facet of the marketing mix is price. Price planning often involves answering such questions as, "What price should we charge?" "Should we go in with a high price, or are we better off starting with a low price?" "What do you think we logically and reasonably can charge for this product?" "Should we change the price, and if so, when and by how much?"

Answers to these questions are never easy to work out. However, the answers chosen are more likely to be correct if they are the result of an organized price-planning system that takes into consideration the company's objectives and the present market conditions. The price-planning system should require that price decisions be based on the company's basic price policy, an understanding of various price strategies, a technique for calculating list prices and, if necessary, a knowledge of how to combat the price-cutting and discounting by competitors.

Price Objectives

Price objectives are essential. Well-stated objectives help dictate management decisions at all levels. If a pricing decision is in line with the company objective, then it has passed the first test of whether it is the right move for a particular situation. The five basic pricing objectives used most frequently by marketers specify that a company's prices should be set to meet the following criteria:

1. *Achieve a target return on investment or a specific net return on gross sales.* Return on investment is the more accurate figure and is the policy used in many of the larger marketing-oriented companies. Return on investment takes into consideration all of the company's resources and how they are used, and measures total costs, both tangible and intangible, against total profits.

2. *Achieve market stabilization.* This is a kind of "live and let live" policy, common in such major industries as oil, steel, and aluminum. The leading company in the industry usually sets the price pattern, and the smaller companies set their prices at or close to the leader's prices. Such pricing is legal and is not considered a form of "price-fixing" or a violation of pricing laws as long as the prices set are both justifiable and reasonable.

If the federal government does not consider a particular price increase in one of these major industries justifiable, it may threaten the industry with serious legal consequences, including a possible takeover of the industry in the interest of the national economy. The industry is forced to roll back its prices, as has happened in the steel industry in recent years.

3. *Meet the competition.* Somewhat similar to the previous point, meeting the competition requires a company to pick prices that closely match those already being charged for similar products in the market. However, it is used in markets that are very competitive as a means of keeping price-cutting or discounting from spreading through such markets and hurting all the marketers involved. Instead of using

price as a competitive weapon, companies in these markets usually try to use the product itself, its promotion, or its distribution, to pick up more sales.

4. *Maintain or improve the company's share of the market.* This is the policy used more frequently than any other, with the single exception of the return-on-investment objective. It is particularly useful for companies that deal in very changeable markets. The policy involves setting prices according to what share of the market the company currently has and what share it wants to have, thereby using price as a major competitive weapon. Thus, it is the complete reverse of the "meet the competition" policy.

5. *Maximize profits.* The way this goal is phrased may sound like profiteering, to the uninitiated. Rather, it is a policy that emphasizes considering all the company's products as part of the total profit picture, and taking a long-range approach to developing profits from those products. A company that plans in terms of maximizing profits may place a low price on a new product in an attempt to win a maximum number of customers for that product, under the assumption that those customers will continue to buy the product over a considerable period of time and thus create maximum profits for the product.

6. *Honor trade discount policies.* Many firms offer prices that differ by the position of the buyer within the channel of distribution. Most companies will treat such customers with special courtesy. Thus, a firm would permit a wholesaler to purchase products at a particularly low price, and another manufacturer at an appropriately low price.

7. *Abide by all pricing legislation.* There are a large number of pricing regulations at the federal and state level. Competitive conditions sometimes make

the consideration of violating such objectives difficult. One example of such laws is "fair trade" laws operating in several states, which make it illegal for firms to sell certain products below a prestated published price. In other words, no discounting.

In 1958, the Brookings Institution in Washington made a study of the pricing objectives of some of the country's leading companies. Here are the policies reportedly used by those companies.

Alcoa: achieve a 20 percent return on investment. *American Can:* maintain share of the market, meet competition, and achieve price stabilization. *Du Pont:* achieve a target return on investment (no figure was given), charge what the traffic will bear, and work toward a maximum return on new products. *Exxon:* achieve a "fair" return (no specific figure was given), maintain a share of the market, and achieve price stabilization.

General Electric: achieve a 20 percent return on investment, achieve price stabilization on nationally advertised products. *Goodyear:* meet the competition, maintain "position" (share of market), achieve price stabilization. *Gulf & Western:* follow price of the most important marketer in each area, maintain share of the market, achieve price stabilization. *International Harvester:* achieve a 10 percent return on investment, achieve a "reasonable" market share. *Johns Manville:* achieve a return on investment greater than last 15-year average, a share of the market not greater than 20 percent, and price stabilization.

PRICING POLICIES

A company's pricing objectives can and should serve as guides for establishing policies.

Initial Price Policies

When a new product is introduced and moves toward its competition, the firm can use two pricing policies: skim-the-cream and penetration pricing.

Skim-the-cream pricing is the strategy involved when a company sets a relatively high initial price on a product. The strategy often is used by companies that want to recover their initial investment quickly, and it can be successful so long as the product is in the early stages of its life cycle and thus is still a distinctive item that hasn't yet been challenged by the competition as in the case where the company has a patent on a new product. The strategy is also used by companies that want to stretch out the growth stage of the product's life cycle and encourage the adoption of the product by those customers who are less sensitive to price.

Skim-the-cream pricing has one important built-in safety factor. If the price is set too high, it can be lowered, which is much easier to do than to raise the initial price of a product. The firm prepares for this development by the tactic of step-down pricing or "cascading" the price downward.

Penetration pricing involves selecting an initial price that is relatively low. Its object is to encourage as many sales as possible. The application of penetration pricing discourages competition, because other companies cannot see how they can bring out a similar product at a similarly low price yet still come away with enough of the market to make sufficient profit.

List Prices

In preparing the actual list price of a new product, a company needs to collect a certain amount of information about the product's market. Often this information already has been worked up during the development of the product. If not, it usually is relatively easy to estimate. Once the company has this information, it must choose which way it wants to figure the list price. A company that introduces a single product at a time, such as an industrial equipment company, usually figures each new price individually. A company that frequently adds many products to its assortment, such as a department store, normally uses a predetermined mathematical formula.

Information Needed. In preparing to set a list price, a company first needs to have an estimate of the probable demand for the product and the price range in which the list price should be set. These usually can be determined by checking the prices and sales of competitive products that are already on the market and by surveying wholesalers, retailers, and prospective customers. If possible, the product should be test-marketed in a few areas to ascertain customer reaction.

Next, how the competition will respond to a particular price range has to be calculated. The single most important factor in any pricing situation is what the competition will do. As discussed earlier, a high initial price often encourages the competition to try to get a piece of the profit, whereas a low initial price sometimes keeps the competition out of the market until the product has become well established.

However, it is wise to remember that there are three kinds of competition to consider: direct competition from competitors with similar products, semidirect competition from companies making dissimilar products that nevertheless serve a similar function, indirect competition from companies making completely different products that serve completely different purposes, but that are competing for the same customer dollar.

For instance, a company that manufactures oil burners is in direct competition with other manufacturers of oil burners. It is in semidirect competition with companies that make gas-heating equipment. In addition, it has indirect competition from companies that handle building materials or landscaping supplies, for a home owner interested in improving the value of his or her property may decide to buy a new oil burner, switch over to gas-heating equipment, or decide to spend the money on adding a new room to the house or improving the grounds.

Finally, a company has to estimate the share of the market it wants the product to win within a specific time period. This is a key estimate. It is the last step before figuring the actual list price. An aggressive firm prepared to capture a large share of the market and with the resources to do so may choose penetration pricing. A company that wants to move relatively slowly, because of the customers it wants to win or a lack of sufficient product to supply a large share of the market, may choose skim-the-cream pricing.

Alternative Pricing Policies

Several types of pricing policy are in use today, which include the single-price policy, the variable-price policy, price-lining, leader pricing, and psychological pricing.

The single-price policy means "one price for every customer, regardless of terms of purchase." It has the great advantage of never being misunderstood by any customer—nor misunderstood by a regulatory agency on the lookout for violations of the Robinson-Patman

Act. On the other hand, the law does allow a company to set *varying prices* that are based on quantity purchased in a single order, delivery time permitted, and other differences in terms of purchases, as long as the policy is used carefully, within the limits proscribed by the law.

Price-lining involves selecting relatively few price lines to handle and concentrating the entire product offering within these price lines. It enables a marketer to concentrate on depth assortment within a few price lines, instead of trying to invest his resources in many price lines. If a marketer picks as his specialty those price lines in which the competition is weak, then price lining can be a tool that concentrates a company's strength against the competition's weakness. That, remember, was an important characteristic of good strategy.

Leader pricing uses the indirect approach. It consists of pricing one or more products attractively low, in an attempt to draw customers who will then remain to buy other products that are priced for a higher return on investment. Leader pricing is a familiar policy of many lower-price and medium-price retail stores, which use a special sale to bring in customers, but expect those customers to be interested in buying other items at regular prices. Sometimes the bargain is even offered as a "loss leader," which in some states is illegal, an item with a price lower than its cost to the store.

Psychological pricing is a matter of making the prices sound right to the target customers. Sometimes it amounts to nothing more than a penny or two. Stores that boast of their bargains usually use odd-figure prices, such as $5.97 or $5.98, whereas outlets that have a prestige image are likely to use round figures,

such as $6.00. Odd-figure prices seem to suggest a special buy, whereas even-figure prices seem more dignified to customers.
Nonprice Tactics. Price-cutting can also be fought with nonprice weapons. Price is only one factor of the marketing mix. There are three other weapons, and a change in any one of them may make that important competitive difference.
  1. *Look at the product itself:* Is there some way it can be changed to make it seem more attractive, more competitive, without changing the amount of resources invested in it such as warranties, money back if dissatisfied, and so forth?
  2. *Look at its promotion:* What about a new promotional theme, a fresh look to advertising, the use of a different medium for advertising than the company has used before? What about premiums, or contests, or giveaways? Anything that can give new excitement to the product and add to its attractiveness in the eyes of the customer can help.
  3. *Look at the distribution system:* Is there another way of packing or shipping the product that would be advantageous? Should different lines and channels of communication be considered? Can warehousing and order processing be made more efficient?

Combating Price-cutting and Discounting
What courses are open to a company when its competition hits the market with heavy discounting, excessive price cuts, and other price-war tactics? Before a company decides to join in the war, it should consider some alternatives. Admittedly, these alternatives are most likely to be successful in a fairly stable market, but even in a highly competitive market, a company faced with hard price competition should check the

possibilities before starting down the road that usually severely damages profits for all companies concerned. Whenever possible, a company should try to meet competition, regardless of its type, not by retailing with price-cutting, for price-cutting decreases profits, but by using all the other possible defenses first.

Last Resorts. When all else fails and a company is faced with the final necessity of joining in with the price-cutters, the move must be handled so that it works in the company's favor. Remember that:

• Off-season reductions hurt less than in-season reductions, because less volume is handled and therefore there is a smaller loss of profit.

• Frequent price cuts on the same brand bring successively smaller gains in that brand's share of the market, as the customer tends to lose confidence in it.

• Temporary price cuts seldom prevent new products from gaining a foothold in the market, for the effect of temporary price cuts is temporary itself.

• Price cuts on new brands are usually more effective than price cuts on established brands, for new brands are not as distinct in the minds of customers as the established brands.

• Price cuts seem to have little power to reverse a decline in the sales trend. When a product's life cycle enters its stage of decline, all that price cuts seem to do is indicate to the customers that the product is on its way out. Some will continue to buy it, but more and more customers will be interested in whatever product is taking the old product's place.

Price-cutting is generally considered seriously in markets with an elastic demand. But if the market is a steady, inelastic one, then price-cutting is usually ineffectual and results only in reduced volume and reduced profit.

PROMOTION STRATEGIES

If a company has the right product at the right price, the next job the marketer has is to make sure that prospective customers learn about it. Promotion is a vital part of the marketing mix, the way the company communicates news about the product and its price to those whom the company hopes will purchase that product at that price.

Practically every kind of product needs some kind of promotion in today's competitive marketplace. If the product is something unique and radically new, that is the promotion message that should be told. If the product is just slightly different from others already on the market, then the company should concentrate on making doubly sure that its promotion message gets through loud and clear so that prospective customers will listen to it rather than to competing messages.

And because those customers are often in a position of being bombarded by messages from various companies all at the same time, it is important that the company tell its product story in the right way, in the right place, at the right time, and through the right combination of promotional ingredients. Working out how, where, and when the product story should be told is a matter of campaign strategy. That is what the promotion of a product is—a campaign. It is not a single effort, but a series of coordinated efforts aimed at achieving a specific objective.

First, a campaign strategy should take into account all the promotional techniques available to the company. These include advertising, direct mail, personal selling, and display, to mention the major promotional methods. Each method also has many possible facets that could be used. Advertising, for instance, could involve the use of newspapers, magazines, radio, television, billboards, or brochures.

Next, the strategy must put the ingredients together in a way that has a synergistic effect. Synergism means that the sum of the whole can be greater than the sum of its parts—that the combined impact of all of the elements of a promotional campaign will be greater than the effect of each element taken separately. For instance, the slogan, "Avis Tries Harder," would have had some impact if it simply had appeared in the company's advertising once or twice or three times, but the company chose to put synergism to work. It uses the theme throughout its promotional campaign, plugging it in a wide variety of media, using it in everything from television advertising to lapel buttons. The result was the total impact of one slogan used in various media has a far larger dollar-for-dollar value than would have been accomplished by using different themes in each type of medium.

The Promotional Mix
Selecting the right mix of promotional ingredients is a tough job, for there are many alternative combinations. There are literally hundreds of possible combinations. More than one combination may be right, but many of them may be wrong. Four factors influence the promotional mix decision: the budget, the nature of the market, the nature of the product, and the stage of its life cycle the product is in.

1. How Big Is the Budget? For a small company with limited funds, the choice of promotional methods may be very simple. A small company may have to depend upon a few sales, a very limited advertising budget, and ample use of direct mail. A large com-

Part Three Strategy 223

pany, however, willing to invest a large sum into promotion, has much more of a job trying to decide which ingredients, in what proportion, would result in the best return on that investment.

The important point is that the company, small or large, looks on the promotional dollar as an investment, the same kind of investment as the cost of the machinery used to produce the item or the cost of the warehouse space used to store the item. In other words, promotional expense should not be considered a current expenditure that must produce immediate results. Instead, it should be considered in terms of its overall contribution to the total profits and the entire life cycle of the product.

2. *What Is the Nature of the Market?* A market can vary according to its geographic size, whether the customers come to the marketer or the marketer must send salesmen to the customers, and in the latter case, whether the customers are clustered together or are scattered. A market can be local, regional, national, or international. A local market might be covered adequately by personal selling or by a moderate amount of advertising in local media, but a combination of promotional inputs certainly would be needed to communicate news about a product through a larger geographic area.

Whether the customers are expected to come to the marketer's headquarters, as they are to a retailer's store, or whether the marketer must send salesmen out to visit customers, as industrial marketers must, also affects the promotional mix. If the customers usually come to an outlet to do their purchasing, then the promotional mix would put its emphasis first on bringing the customers into the outlet and then on convincing them to buy. If the company has to go to its customers, the promotional emphasis has to be on getting the customer to allow a salesman to visit him.

Where customers who have to be visited are located also makes a difference. If customers are clustered together in a small area, then each salesman can handle several calls a day. If the customers are scattered, then a salesman may have a day's traveling to do between one call and the next. In the first case, a company might put its promotional emphasis on personal selling. In the second case, a company might align an equal amount of media advertising and mail promotion with the personal selling effort.

3. *What Is the Nature of the Product?* If the product is a branded convenience item, the promotional mix probably will place its emphasis on advertising, and the amount spent on personal selling will be just enough to hire clerks to take orders. If the product is a specialty or a service, then the job of personal selling takes on major importance.

4. *What Stage of Its Life Cycle Is the Product In?* The promotional effort is most important in the earlier stages of the product's life cycle. The emphasis in the introduction and growth stages is on interesting the customer in the product, and any means that encourages interest should get preference in the promotional mix. In the industrial market, for example, both personal selling and trade advertising are very important, whereas in the consumer market, advertising in mass media is essential. Promotional efforts during the maturity phase usually concentrate on those ingredients that tell the customer where the item can be purchased and that encourage purchase from one marketer over another. As the product reaches the stage of decline, the promotional effort usually shrinks down

to only a minimum of advertising, and the personal selling effort is switched over to newer products.

No Formal Guidelines. There are no formal guidelines for putting together a winning promotional mix. The factors outlined above are far too varied. Nor is any combination the single right combination, for how well a combination succeeds usually depends completely on the ability of the company handling the promotion and the sales results obtained specifically from the promotion. However, there are some informal, rather broad guidelines that should be considered. They are neither laws nor rules, but simply observations based on past promotional experience.

Personal Selling. This is the promotional tool that should be of major importance in situations where the company's funds are limited, where there is a concentrated grouping of customers, or where the nature of the product requires demonstration or personal counseling. Automobiles and expensive industrial equipment, for instance, require an emphasis on the personal selling technique or promotion, as do all kinds of services that require explanation rather than product identification.

Advertising. This should be the main ingredient in promotional mixes for a product that has wide appeal and is expected to have broad acceptance, such as many kinds of consumer products. No amount of advertising can sell a product that people do not want to buy. However, when people are interested in purchasing an item, informative advertising can help them decide which of several competing products they will choose.

Advertising is also important in helping to back up the personal selling method by preselling the product prior to salesmen's calls. Its role in the industrial market is slightly different from its role in the consumer goods market. Advertising in the industrial market is usually intended to convince a prospective customer to permit a salesman to visit him, while advertising in the consumer market is intended to bring the customer into the sales outlet.

Other Sales Promotion Methods. Other techniques, such as display, sampling, catalogs, coupons, and contests, all have their special uses. Display is very important in most retail stores, particularly for products whose characteristics can be judged easily at the point of purchase or products that are often bought on impulse. Sampling, coupons, and contests are usually used to encourage customer interest in a product. Catalogs are very useful for selling products whose purchase does not require a personal inspection by the customer.

DISTRIBUTION STRATEGIES

A few years ago, O. M. Scott and Sons, manufacturers of lawn products, included this statement in the company's annual report:

There are three basic components for marketing success. 1. Products must be those that build leadership because they satisfy the customer; 2. advertising and promotion must be the type that translates this leadership into buying interest; and 3. products must be available at the time and place required to convert favorable attitudes into sales.

Many companies, however, take the first two components far more seriously than the third. They feel they can control the product and its promotion, but that its distribution is uncontrollable, a matter of what has always been done rather than what could be done. As a result, too many companies simply follow the distribution patterns that have become traditional in their particular markets and never plan ways that might improve their distribution methods.

For example, some of the large meat-packaging companies remained with branches located along the railroad tracks long after refrigerated trucks, centralized buying, regional warehouses, and prepackaged meats had changed the entire logistical nature of the meat-packing business. Other marketers, when asked about possible distribution changes, will answer, "Why should we bypass the wholesaler?" "Think of Air Express, when we're practically sitting on top of the railroad tracks?" "Establish regional warehouses, with what that would do to our costs?" What was good enough yesterday is good enough for today and tomorrow, according to these marketers.

On the other hand, consider those companies that have broken with traditional patterns. Think of all the merchandise that is now being prepackaged and shipped directly from manufacturer to retail store, often with the store's price labels already attached. Think of the communication systems that link store, warehouse, and manufacturer so closely that order processing is almost automatic—and think of the warehouses themselves, with conveyor systems that load and unload distant bins and shelves at the touch of a switch.

Chiquita bananas are a good example of a product whose sales in the 1950s and 1960s were boosted by a streamlined method of distribution. The product's promotion concentrated on the way the bananas are packed and shipped so that the fruit arrives on the market in good condition. The company's success proved that the methods by which goods move through the channels of distribution can be important competitive weapons as well as operational necessities.

Cost of Distribution
Distribution expense, like that involved in the product production and promotion, should be considered a long-term investment in the total profitability of the product during its entire life cycle. Distribution costs are generally divided into three groups:
- *Operating costs* are the costs of actually moving the goods, including packing, transportation, loading, and unloading.
- *Possession costs* connected with warehousing, insurance, and inventory control.
- *Service costs* are the costs of processing orders, as well as the penalties paid when special shipments are required to get the product to the right place at the right time.

Spending Strategy
Ideally, the real breakthrough in planning for improved distribution lies in finding a way of circumventing the existing patterns and discovering a route not previously used that is faster or more economical. This is using the technique of the indirect approach. It involves abandoning the expected and trying the unexpected. Here are some of the ways a company might explore.

Change mode of transportation. Consider a faster way to get the product to the market, if that will result in its getting there at the time when it can be sold most profitably. Switching from rail to air transportation would increase transportation costs, but it could reduce warehouse costs and also could enable the marketer to sell the product at a higher price because of its timeliness on the market. Or consider a slower, less expensive way, if the product is one whose time of arrival isn't crucial. The water route is used profitably by many shippers of bulky, but relatively inexpensive, products.

Get more systems and records help. This could mean hiring more clerical workers, or it could require installing more advanced equipment. Either solution would entail increased initial expense, but would also mean that more orders could be processed with greater speed and accuracy. The result would be a reduced need for warehouse inventory and probably fewer lost sales because of processing mistakes. It might also mean that the company could use the time saved to take advantage of slower but cheaper ways of getting the product to the market or customer.

Combine shipments. Most transportation companies charge cheaper rates for bulk shipments, and many manufacturers and wholesalers charge less per unit for large quantity orders. Take advantage of such bargains whenever possible. Hold off before placing an order or making a shipment until the quantity involved is large enough to earn a discount.

Increase the number of warehouses. This is a major logistics decision. The strategy could enable a company to get shipments to outlets and orders to customers more quickly, thus making the product more attractive. In contrast, it may be possible to give the company added geographical distribution speed without the need for building new warehouses by establishing a new middleman link in the distribution network.

Automate the handling of the product. Transportation and warehouse systems have the capacity to be partially or almost completely automated, enabling the product to be loaded, transported, unloaded, and stored more quickly, more efficiently, using fewer personnel.

Automate the inventory system. This change involves an expense again, but it can pay off in better control of the product and often a lowered level of required inventory (and the less money a company has to keep tied up in inventory, the more money it has available for investment in other segments of the operation). Genesco, for example, has an inventory of about 1.5 million pairs of shoes, and its automated distribution system coordinates the output of 17 manufacturing facilities and handles the shipment of about 40,000 pairs of shoes daily to its 1,500 retail outlets.

THE MARKETPLACE

A marketer who is putting a plan into action must consider all four facets of his marketing mix in terms of the particular market he wants to penetrate or maintain. Not all markets are the same, nor are all customer groups alike. Each market, each group of customers requires a different marketing approach. Here are two ways of planning for a share of a particular kind of market, one that concentrates on winning a single segment of a large market and one that explores the international possibilities of markets.

Market Segmentation
Market segmentation divides a market according to separate characteristics. For example, a manufacturer of men's slacks might plan in terms of getting a share

of the total market or in terms of winning a larger share of a particular submarket or market segment. Such a segment might involve teenage mod groups, college men, businessmen, or retired men. It might involve a specific geographic area, such as California or New England, Long Island or Harlem. It might involve end use, such as golf slacks, ski pants, walking shorts, or Bermuda shorts.

The decision to concentrate on a particular segment of a market can take a wise bit of planning. Market segmentation is similar to the basic planning strategy of concentration of resources. It is an attempt to present a great deal to a selected few, instead of trying to offer a little to everyone.

Advantages. When a businessman decides to concentrate the company's money, manpower, and materials on a market segment, if he can find a profitable segment he has the following advantages:

- Head-on collisions with the competition are few, for the marketer avoids any segment that already looks saturated.
- Marketing resources can be allocated where there is the chance of the best possible return on investment.
- Because competition is not intense, competition strictly on the basis of price is usually avoided and product differences can become the major sales emphasis.
- The marketer is less vulnerable to whatever competition does exist or that appears, for his resources are concentrated in strength in the segment where research showed that they would be most effective.

Choosing the Segment. However, not all markets can be segmented. To determine whether a market can be segmented, and whether it should be considered a possibility, a marketer should do research that answers the following questions:

- Can the market be measured? Can enough information be gathered about it to measure its potential?
- Is the market accessible? Can the company penetrate the market economically and profitably?
- Is the market substantial? Is it worth going after? Will the segment generate enough sales and profits to justify the effort?

How to Proceed. The next step is to determine the basis for segmenting the market. Here are the most frequently used methods.

By product. This is the simplest and most common means of segmenting a market. The television market, for instance, can be segmented into black-and-white sets and color sets, into portable sets and console sets. The radio market can be segmented into AM and FM reception, into console sets, clock radio sets, and portable sets. Books can be divided into hardback and softcover, fiction and nonfiction, general interest and special interest. In other words, when a market is divided by product segments, it is considered in terms of how or why the customer uses the product, not the techniques used by the marketer to sell that product.

By demographics. This method segments a market by age, income, sex, geographical location, race, marital status, education, among other statistics. It may be the most widely used method of segmenting a market simply because so much statistical information is available. It is the method that prompts a company to determine that the product it sells is sold primarily to married couples, over 50 years of age, who live in rural areas of the Northeast. The company then concentrates its sales efforts only on those who meet these qualifications.

By psychographics. This method is difficult to work out, and is still a relatively new technique in segmentation, yet it is a very valid kind of planning when used properly. Customer groups are divided by attitudes, behavior, values, tastes, and lifestyles. Retailers frequently use this method in determining the target group of customers they want to serve. For instance, a store may stock conservative, higher-priced sports clothes, with a very particular segment of affluent, leisure-minded target customers in mind.

By special consumer interests. The final method is to divide markets by the special interests they contain. Sports clothes, for instance, represent a broad market, but golf, tennis, or skiing clothes are a much narrower segment. In order to interest customers who have these special hobbies even more, the marketer may add considerable depth to the assortment he offers. For instance, he may offer not only golf clothes but also golf clubs, golf balls, golfing novelties, perhaps even golf lessons.

A Key to Growth. Market segmentation as a planning strategy is not only a practical way of looking for more profitable areas in which to do business, it has also become almost a necessity because of the rapid and diverse changes in customer interests and desires. It is an important key for marketing growth, and is the way much marketing business will be done in the future. Specialization, in markets as well as products, is becoming increasingly important.

Here are some of the ways in which the increasing emphasis on automation will provide markets for businessmen who are willing and able to specialize in one or more facets of the technology. Currency and checks are being replaced by automatically controlled credit systems. Consumer and business purchases will someday be made from home or office through displays of goods presented on private color television. Medical diagnosis and treatment will take place in computer-monitored "treatment capsules." Students will be taught by programmed learning units that feed on computer-stored knowledge banks. Doesn't this alone suggest a variety of market segments that a marketer might exploit profitably?

International Markets
One of the last great frontiers for business expansion is the international marketplace. When domestic markets become saturated and competitive pricing and promotion at home cut into volume and profits, it is possible to sustain volume and expand opportunities by probing foreign markets. At present, there are approximately 20,000 U.S. companies doing business in overseas markets. Some just happened to get into international trade; others deliberately planned their international expansion.

Getting Started. The key planning approach is to consider each foreign market as a new opportunity for expansion, not as a dumping ground for surplus production that cannot be sold domestically. Each country should be studied separately. Check the policies of each with regard to American companies. Find out what competition exists in the particular market. Estimate the degree of acceptance or resistance a product is likely to meet, both from governmental regulations and from the point of view of potential customers.

A great deal of information on international markets is available, which can be obtained without going abroad. Talk to the international trade specialists at the commercial banks; they know about foreign credit, international market conditions, foreign methods of collection, foreign taxes. Contact advertising agencies that have had experience in foreign markets. See the various trade representatives attached to the foreign embassies. Check through the numerous publications and other sources of information at the United Nations and other international organizations.

Perhaps the best source of advice and information is the U.S. Department of Commerce:

We in the United States Department of Commerce are enthused with an international marketing package . . . the Global Marketing Program! U.S. manufacturers in the Global Marketing industries are being offered a marketing package that includes specialized reports of research, sales leads, and an unprecedented level of on-the-scene professional events and sales opportunities geared to their product lines. One aim of this effort is to make U.S. companies feel as much at home in dealing with foreign markets as with their more familiar domestic customers.

Entering the Foreign Market. There are four ways to get into a foreign market—exporting, establishing foreign sales offices or branches, arranging a licensing agreement, and setting up a joint venture. Exporting is by far the easiest and least expensive way to pick up foreign sales. Exporting can be done indirectly through export/import companies and commission houses, or it can be handled directly by finding and supplying customers abroad. The direct method enables a marketer to maintain more control over the selling operation, but it also means that the marketer must handle the export packing, documentation, international shipping and insurance, customs clearance, and credit and exchange arrangements that otherwise would be handled by an intermediary agent.

Establishing sales offices in foreign countries centralizes the company's operations in that country and eliminates the need for international agents. Many American companies often set up warehouses and manufacturing facilities abroad. For years Coca-Cola has shipped syrup in bulk to foreign markets, where it is then diluted and bottled. Appliance manufacturers often find it more economical to ship the parts of large appliances, such as refrigerators, for assembly at their foreign plants, rather than ship the finished appliance.

The advantage of a licensing arrangement is that domestic companies can arrange to have products manufactured and marketed abroad without financial outlay. The American company permits a foreign company to use the patents, processes, or trademarks of the domestic company in return for royalties on all sales of the product.

A joint venture involves an exchange of stock or a merger between the U.S. company and an existing foreign firm. It can make for quicker acceptance of the new "local" company abroad on the basis of the market already built by the foreign company. Thus a good deal of the development work that would ordinarily have to be done in a new market can be avoided. But before you sign any such agreement, make sure that the foreign company's general reputation, its manufacturing and sales capabilities, and its financial strength, have been thoroughly checked out.

Global Marketing. Managers abroad should get to know and understand the culture, tastes, mores, and behavior acceptable to that country's businessmen. For instance, business is seldom discussed during lunch in France, while in Greece, one is expected to drink cup after cup of coffee while talking business.

A manager should know the foreign market and its people. When the McGraw-Hill Book Company opened a Canadian subsidiary, it did not simply take an existing American textbook, change a few words in the text to "Canadianize" it, and put a new cover on it. Instead, it worked out a line of books written by, for, and about Canadians. The other side of the picture is illustrated by an American poultry co-op that decided to export turkeys to Europe. It financed an impressive and expensive promotional campaign in Germany about the delights of roasted turkeys—but nobody realized that the average overgrown American turkey would not fit into the average small-size oven found in most German kitchens.

It is a good practice to hire operating personnel from the people who live in the foreign country. It's fine to have managers who are U.S. citizens, if that is the company's practice, but a marketing effort is most successful when prospective customers feel comfortable—and people feel more at ease with their own. Success in global marketing, like success in every aspect of marketing, is a matter of wise, deliberate planning.

Appendix
Brain-Teasers
Thought-Provokers

We tend to understand things best if we have to invest some brainpower in them. The learning process appears to work best when we are actively and mentally involved. To help you in considering the marketing executive's role and problems, we have provided the following exercises in the form of brain-teasers, and thought-provokers.

The brain-teasers should help you assess your marketing knowledge. The thought-provokers should place you in the position of considering issues and problems as the marketing executive does. The material is so organized that you can check your conclusions against the text with considerable ease.

BRAIN-TEASERS

Chapter 1

Agree Disagree

1. Peter Drucker considered marketing the unique function of the business—without it, a company really wouldn't be operating with the proper objectives. ____ ____
2. Built into the systems approach is the mutual stake in survival theory, which assumes that all aspects of a system are necessary. ____ ____
3. Marketing primarily involves activities that occur after the time of production. ____ ____
4. There is little reason to be concerned with the shift in consumer demand as long as production lines can handle a potential increase. ____ ____
5. The role of the marketer has increased primarily to fill a vacuum within the organizational structure; it will be reduced once this void has been filled. ____ ____

Chapter 2

Agree Disagree

1. The need for business planning is important, regardless of the economic system. ____ ____
2. Success in a competitive market depends on the ability of a marketer to anticipate technological change and to plan accordingly. ____ ____
3. If enough attention is focused on efficient production methods, there is less need for marketing strategy. ____ ____
4. One of the aims of marketing strategy is to assure that the products offered are acceptable in the target area. ____ ____
5. The primary aim of marketing is to get the product accepted by customers. Profits are of secondary interest. ____ ____

6. *Marketing orientation* adapts production to the needs of the customer; *product orientation* places importance on giving the customer what can be made efficiently. _____ _____

Chapter 3

                                                                                 Agree   Disagree

1. A primary benefit of planning is that money, manpower, and materials can be accounted for easily. _____ _____
2. Planning assures that each manager is working toward achieving the objectives of his or her particular department. _____ _____
3. The proper relationship between policy and objectives makes objectives an instrument of policy. _____ _____
4. Reducing the number of organizational levels between the sales force and top management allows for more flexible planning activity. _____ _____
5. Marketing orientation tends to have only a moderate impact on a company's stated policies and objectives. _____ _____
6. When company policies come under review only the marketing team should have the greatest impact on the issues, under the marketing concept. _____ _____

Chapter 4

                                                                                  Agree   Disagree

1. The quality of the data is only as useful as the reliability of its source. _____ _____
2. A market research project most often begins with the acquisition of primary data followed by a review of secondary data. _____ _____

Appendix   Brain-Teasers

3. A review of internal records reveals past buying patterns of existing products, and is useful in projecting the product life cycle of a new product. ____ ____
4. The ideal use of a structured questionnaire is in person-to-person interviews. ____ ____
5. Observation is an economical method of collecting primary data. ____ ____
6. Selecting test markets is an application of collecting data through experimentation. ____ ____
7. Data may be classified into three categories: hard facts, opinions, and behavior. ____ ____
8. In the normal course of operating a business, most marketers collect both control data and planning data. ____ ____
9. It is generally wise to collect whatever internal market information is available. ____ ____
10. The prime advantage of a marketing information system is that it eliminates most of the human factor from the decision making process. ____ ____

Chapter 5

Agree   Disagree

1. Small samplings of sales are generally sufficient in providing reliable data for use as trend indicators. ____ ____
2. Market share analysis is most useful in markets that reflect frequent shifting of product preference and customer demand. ____ ____
3. Of the various methods of analysis discussed in Chapter 5, a functional cost analysis is the best method of calculating profit and cost. ____ ____
4. Break-even analysis examines only the relationships for cost, profit, and sales volume. ____ ____

5. If profits for a new product are projected high enough, there is no need to consider return on investment. ____ ____
6. The primary purpose of ROI analysis is that it is an effective means for helping managers determine the best use of capital. ____ ____
7. The advantage of using various forms of market analysis lies in its accurate, unbiased, cause-and-effect relationships. ____ ____
8. One of the outcomes of good planning is a reliable forecast. ____ ____
9. A sales force can generally be counted on to give accurate sales forecasts because of its close relationship with the customer. ____ ____
10. Forecasting is largely guesswork. ____ ____

Chapter 6

Agree *Disagree*

1. Decision making is always a difficult task. Decisions that cover a broad spectrum of instances are harder to make than one-time decisions because they continue to recur. ____ ____
2. Role-playing is a form of model-building in which members of the marketing department "act out" roles in order to get a feel for how certain decisions might work out in actual practice. Since the approach is somewhat contrived, it has very limited value within the marketing area. ____ ____
3. Game theory works on the assumption that history will repeat itself. For example, if every time one firm has placed a full-page ad in the local newspaper a second firm has reacted by placing a

Appendix Brain-Teasers

similar ad, chances are that the same type of behavior will occur again. For this reason, game theory has particular advantages in certain frames of reference.

4. The fewer the number of competitors in an industry, the greater the risk to the company that acts independently of other companies in the same industry.

5. Under the "minimax" theory decision makers look on the bright side of things. They assume that their competitors will not be successful and that their own position will hold firm.

6. The particular advantage of a decision tree is that it forces managers to analyze the consequences of their decisions.

7. PERT charts are scheduling devices that serve primarily to identify excess time requirements within marketing programs.

8. Because one of the roles of marketing management is to predict and project into the future, it is important that the marketing plan be a step ahead of the corporate plan.

9. The marketing plan functions as a static schedule on which to place decision alternatives. This being the case, although the basic plan itself does not have a tendency to change once it has been established, it may be modified frequently in small details.

10. Marketing decision making tends to be more responsive within the firm than is perhaps true in other areas of the firm, primarily because a large number of such decisions have to do with volatile subjects, such as market conditions, shifts in demand, shifts in competitive conditions, and the like.

Chapter 7

	Agree	Disagree
1. One of the functions of a marketing plan is to provide a framework for combining the ingredients of the marketing mix into an effective marketing strategy.	_____	_____
2. Using an unorthodox strategy in a marketing effort is too risky and is generally not recommended for smaller companies.	_____	_____
3. The primary function of a computer in marketing is to relieve the marketing manager of having to make final decisions.	_____	_____
4. Only the controllable marketing factors are incorporated into the marketing plan. Since a marketer generally can't alter the "uncontrollables" of the marketplace, there is no need to factor them into the plan.	_____	_____
5. A marketing strategy that has failed in one situation may be used successfully when applied to another situation.	_____	_____
6. A company with a good sales force, a big advertising budget, and expanding manufacturing capabilities can be assured of success in the marketplace over direct competitors that have a smaller sales force, a smaller advertising budget, and a relatively limited manufacturing capability.	_____	_____
7. The nature of business pits one company against another in a product-against-product battle.	_____	_____
8. Grand strategy is the overall corporate objective; working strategy is putting the individual marketing plan into action.	_____	_____
9. Speed, rather than slow deliberate action, is essential to implementation of a successful marketing strategy.	_____	_____

Appendix Brain-Teasers

10. Finding the marketing center of gravity is the key to unbalancing the competition. _____ _____

Chapter 8

                                                           Agree Disagree

1. The ingredients of the marketing mix consist of product, price, promotion, and distribution. However, in planning it is necessary only to consider one of these. _____ _____
2. The statistics that seven out of ten new products do not become commercial successes would suggest that the marketer deemphasize new product development and stick with the existing profitable lines. _____ _____
3. One of the basic product strategies is to beat the competitive product. _____ _____
4. The most concentrated sales effort in the life cycle of a successful new product takes place during the introductory stage. _____ _____
5. The first stage in a pricing decision should be to complement company pricing policy before considering the competitive or customer pricing situations. _____ _____
6. Where market pricing is the industry practice for determining selling price, other elements of the marketing mix must be used to seek out additional profits. Price no longer is a key factor. _____ _____
7. Achieving the effect of synergism in a promotion campaign is really another way of saying that you have created an effective promotion mix. _____ _____
8. With the expansion of freight-handling capabilities by the airlines industry in recent years, the use of water transportation for hauling freight is becoming obsolete. _____ _____

9. A primary criterion for establishing a market segment is that initially there is less concern with competition. _____ _____
10. International business is generally best suited to the larger companies; smaller companies are unable to compete because of all the problems of international trade. _____ _____

THOUGHT-PROVOKERS

Chapter 1

1. Consider the magnitude of Peter Drucker's comment about marketing as the unique function of the business. Can you think of a business that has no marketing department or function? What would such a firm do? What would be its objectives? Should it be run like a business? If not, how?
2. Considering that marketing is only one of about five or six major business functions, why is so much attention being placed on it? Is there any reason to assume there will be a shift in corporate emphasis? What might cause such a shift?
3. Taking your company, or one with which you are familiar, try to trace the stages of the introduction of a new product. Try to determine if there was any communication with consumers.
4. What factors come to mind that might focus even more attention on the marketing structure?
5. Prepare a list of questions on marketing decisions with which you tend to disagree. Then examine why the decisions or policies are as they are.

Chapter 2

1. Planning takes place in businesses within both a capitalistic system and a controlled economic system. Compare businesses within each type of system and indicate the extent to which planning is needed in such business activities as administration, production, finances, and marketing.
2. From the beginning of the Industrial Revolution in the late 1700s to the end of World War II in the mid-1940s, emphasis on methods of production, rather than needs of the market, formed the basis for the economic growth of the United States. What are the reasons why the production approach was successful for that period, and why couldn't that approach continue successfully after World War II?
3. One of today's serious marketing problems is the shortened product life cycle—in some cases from years to months. Examine various products and product lines and give specific examples of these problems. Explain ways in which planning could alter the product life cycle.

4. Statistics show that approximately seven out of every ten new products do not succeed in the marketplace. Yet, the mainstay of every business is new-product development. What are some of the factors that must be considered if market resistance is to be overcome?
5. An executive in a Midwest industrial firm was quoted as saying, "A marketing manager is a brand new title for the same old job of sales manager." How would you react to the statement? Inquire about a company that has a marketing manager and one that has a sales manager. Contrast the duties of each position.
6. Consider examples of companies that have adapted the marketing concept. Then contrast the changes that have been made with companies that, in your opinion, are still fixed in a production orientation.

Chapter 3

1. List the six benefits of planning, giving specific examples of how each can be applied to your work unit or organization.
2. What are the policies and objectives of your company? If they are not generally known, interview the appropriate executive. Classify the objectives as long-range and short-range, and rate them according to the criteria listed in Chapter 3.
3. Compare the way the production function would differ if its policies and objectives were organized in terms of the production approach and the marketing concept.

Chapter 4

1. A company you are working for has an idea for a new type of camera. The developers think it can sell for less than $25; yet it has features normally found in cameras at twice the price. What possible sources of secondary information would you need, and what type of data would you expect to obtain from each source?
2. Using the above product, develop two questionnaires: one for use in a mail survey, the other in a personal interview.
3. What people say they will buy or how they will buy is often different from what they actually go out and buy. Select one

aspect of buying behavior gathered through your secondary or primary camera research. Using the observation method at the point of purchase, check the accuracy of your original information. Also determine what additional data may be obtained through observation.
4. Select a product and describe an experiment to alter the product or its packaging. Use the guidelines described in the text to establish a control.
5. Examine the data available in your company or in one with which you are familiar, and classify the data generated as control data or planning data. Indicate how often the data is made available, and then rank the information in terms of its usefulness.
6. On the basis of your findings, propose a detailed data collection system that you think you can "sell" to management. With your recommendation, submit the estimated costs of such a system.

Chapter 5

1. Preparing data for use in the marketing plan is as vital as the task of collecting the data. Identify someone in your organization who you think has the qualifications suitable for preparing data. What criteria did you use in making your choice?
2. Many firms do specialized forms of market analysis. Identify companies that do market analysis and categorize them by specialty.
3. An example is given in the text of a shoe store manager's need to determine the future inventory. What classifications of data would be needed to yield the necessary information? What kinds of tables, charts, or graphs would the manager make use of?
4. Consider a large department store in your area and a manufacturer of power tools for the home market with national sales distribution in terms of the classifications each company would need to establish for a good sales analysis. What similarities do you find? What are the differences?
5. Many industries, such as automobiles, cigarettes, and soft drinks, publish annual share-of-market figures. Examine at least three industries that publish such information. In your opinion, how significant are the rankings of companies by

share of market? Is there a relationship between share of market and profitability?
6. A manufacturer of a new type of radio has projected first-year sales at 10,000 units at a unit selling price of $18.00. Variable costs are $100,000 and fixed costs are $35,000. Construct a break-even chart. Then show the break-even point at a volume of 15,000 units. Selling price and fixed costs remain the same. However, the variable costs have increased to $110,000.

Chapter 6

1. Assume that you operate an appliance store and that you are making a marketing plan for the coming year. You have no data on hand other than your own past sales, so you are interested in obtaining information for your plan. Identify ten sources of appropriate data for such a task.
2. You are planning to open a store in a nearby community, but you have paid little attention to the competitive situation within the community, and you are not quite sure about the size and shape of the trading area. You need this information for your marketing plan. How would you go about obtaining it?
3. Using the activity time periods shown in the following chart, construct (a) a PERT network and (b) the critical path.

Event	Weeks	Event	Weeks
1-2	3	2-4	4
1-3	2	3-5	3
2-5	4	4-7	2
6-8	3	4-6	4
8-9	4	5-6	4
7-8	2	7-9	3
6-7	3	8-9	2

4. Build an action deadline for women's winter fashion coats from the points of view of the wholesaler and the retailer. Select an industry of interest to you. Check some current economic indicators of the type found in the Sunday business press. Then attempt to estimate the probable impact of these indicators on sales in your chosen industry over the next month or two.

Appendix Thought-Provokers

5. What is the current trend of the GNP? The current growth rate of the economy? The current rate of inflation? What is the general trend of the economy?
6. Identify some examples of the use of minimax, maximax, and maximin from newspapers, store visits, and other sources.
7. Which of the following items belong in a marketing plan? Which fall into a corporate plan?
   a. Corporate objectives
   b. New facilities plans
   c. Competitive profitability estimates
   d. Forecast of sales
   e. Economic forecast
   f. Estimates of corporation profits
   g. Workforce estimates
   h. Implementation and control statement
8. You operate a business that is considering expansion. The projections are that company sales will rise by 50 percent over the next five years, and your plant is operating at full capacity. Sales have risen at the rate of 5 percent over each of the past three years. Money is tight, so you don't want to waste it. Construction costs are high, and your production costs will rise by 10 percent if you build a plant that will produce twice as much product, and 5 percent if you build a plant that will produce only 20 percent more product. The current profit is at the rate of 18 percent. You estimate that sales will at least keep up with previous sales, and there is a 70 percent chance that sales will reach the new higher rate. What should you do? Try to answer this problem by constructing a decision tree.
9. Construct a set of the product and promotion questions you would want answered in an annual marketing plan.
10. Compare last year's economic projections with what really happened.

Chapter 7

1. Explain the relationship between corporate goals and marketing objectives for a company organized under the marketing concept.
2. The total market environment for any product comprises controllable and uncontrollable factors. Select a bank, a cigarette manufacturer, an automobile maker, and a tax preparation ser-

vice. Identify the controllables and uncontrollables for each company, and then compare the similarities and differences.
3. One of the criticisms lodged against American automobile manufacturers was the amount of time it took them to move into the small-car market. What element of strategy fell under criticism? Was the wait-and-see approach a good idea or a blunder? Discuss alternative courses of action.
4. Chapter 5 discusses examples of companies that use the indirect approach to circumvent the traditional rules of marketing a product. Examine various consumer and industrial products, and select one from each category that would represent the use of an indirect strategy in product, price, promotion, or distribution.
5. Select a company from the above activity and describe its "shape" using the guidelines set forth in the text. Also try to identify its marketing center of gravity.
6. The strategy of two zones of activity is used by many companies today, especially among the larger organizations. Describe how General Electric, American Telephone & Telegraph, General Motors, or any other large company demonstrates the strategy.

Chapter 8

1. Examine the organizational structures of a consumer goods manufacturer and an industrial products manufacturer. Draw up organizational charts that illustrate how each company could start or improve its new-product development capability. What, if any, are the differences between the two charts? Explain.
2. Select several new products from the two companies examined in the previous activity, and determine which of the three basic product strategies was used to launch the product.
3. Research a product of your choice and plot its life-cycle curve.
4. Using the four methods described in the text for extending the life of a product, select three products and list examples of how each stage could be prolonged.
5. Examine several brands of a product, such as men's shirts. Compare which pricing policies seem to be inherent to the companies as well as the industry. Also determine which pric-

Appendix Thought-Provokers 255

ing strategy—skim-the-cream or penetration pricing—is likely to be used if the companies were to add new products to their existing lines.
6. Select a product you are familiar with and for which statistics are readily available to you. Determine the optimum selling price, and using the break-even analysis, determine the probable quantity that will have to be sold before a profit can be made.
7. Avis was given as an example of the synergistic effect achieved in developing a promotion effort. Give other examples of synergy in promotion campaigns.
8. Given the methods in the text for determining how to segment a market, prepare a segmentation study for a product or service of your choice.

Index

action: deadlines, 141; sequence in collecting data, 75–79
advertising: effectiveness and strategy, 173; promotion and, 225–226
advertising agencies, as sources of secondary data, 60–61
Alcoa Corp., 214
Alderson, Wroe, 4–5
allocation models, 135
alternatives, planning in emphasizing, 35
American Can Co., 214
authority, mid-management, 49–50

break-even analysis, 102, 106–108
Brookings Institution, 63, 214
budget(s): characteristics, 116; marketing planning and, 135–136, 138–139; promotion and, 222–223; types of, 115–116, 136–138
budgeting, 113–116
Business Periodicals Index, 59

capital expenditures, projected, 137–138
cash flow: discounting, 104–106; projected, 137
chain of command, 45–46; reevaluating, 46–50
change: acceptance of, 20–21; resistance to, 21–22; technology and, 13–18
Civil War, growth of technology and, 14
Columbia Record Club, 179
competition: and price objectives, 212–213; and strategy, 164–166, 194–196; unbalancing, 180–186
Conference Board, 63
consumer: attitudes and tastes, 160–162; interests, market segmentation by, 233–234
controls, planning in facilitating, 37
corporate structure, marketing concept and, 29
cost: of distribution, 228; functional, 100; of goods sold, estimate of, 138; vs. sale price, 22

256

Index 257

critical path method (CPM), and PERT, 144–149
customer: increasing importance, 15–18; strategy and environment of, 196–197

Daniel Starch and Staff, 62
data, 53–54; checking for accuracy, 89; classifying, 90–92; deadlines, 141; performance analysis, 93–102; preparation of, 87–88; primary, 55–56, 65–75; secondary, 54–55, 56–65; sequence of action in collecting, 75–79; utilization, 92–93
data analysis pitfalls, 108–110
data collection: sequence, 75–79; systems, 79–85
decision making, 123–129
decision theory: decision criteria in, 130–132; decision tree, 132–134; probability theory, 129–130
decision tree, 132–134
demand analysis, 121–123
demographics, market segmentation by, 232–233
discounting, combating, 219–220
distribution: analysis, 97–98; and channel management, 158–160; strategies, 226–230
diversification, and market orientation, 26–28
Dodge, F. W., Corp., 62
Drucker, Peter F., 3
Du Pont Co., 209–210
Dutch Cleanser, and acceptance of change, 20–21

economic conditions and strategy, 162–163
elastic demand, 122
environment, customer, and strategy, 196–197
environmental forces, on marketplace, 18–22
equipment, MIS, 83–84
exception reporting, 80
expense(s): income and, projected, 136; strategy and, 171; types, 114–115
experimentation, in primary data collection, 73–75
external data, 54
external sources of secondary data, 58–65
Exxon Corp., 214

financial condition, projected, 136
financial management, questions on marketing role, 6
financial resources, identifying position of, 134–135
financial structure, 102–103
fixed costs, 114
flowcharts, 143–144
Ford, Henry, 22
forecasting, 111–113
foundations, as secondary data sources, 63
functional cost, 100

Gallup and Robinson, 62
game theory, and decision making, 127–129
General Electric Co., 25, 214
Genesco, Inc., 230
Goodyear Tire & Rubber Co., 214

government, as secondary data source, 64–65
Great Depression, and switch in market emphasis, 14–15
gross margin, 99
Gulf Oil Corp., 214

headquarters staff, questions on marketing role, 6–7
human factor, in data collection, 84–85

income and expenses, projected, 136
Industrial Revolution, growth of technology and, 13–14
inelastic demand, 122
initiative, encouraging, 43
innovations, environment and, 18–22
internal data, 54
internal sources of secondary data, 57–58
International Harvester Co., 214
international markets, 234–237
interview, in primary data collection, 67–68
inventory, automated system, 230
inventory and materials budget, 138

Johns Manville Corp., 214

Korvette's, 207

leader pricing, 218
legal constraints, strategy and, 163–164
L'EGGS, 177–178, 182–183

Levitt, Theodore, 12
libraries, as sources of secondary data, 59–60
list price, 215–217
long-range marketing plans, 119
long-term objectives, 40

management, interdependence, 7
managers, survival problems, 4–5
market: international, 234–237; shape, 189–191
marketing: center of gravity, 185–186; defined, 8; post–World War II glutting, 16–17
marketing concept, 22–23; market orientation and, 24–28; organizing for, 28–30; production orientation and, 23–24
Marketing Information Guide, 60
marketing information system (MIS), 83
marketing management, coordinated, 12
marketing mix, 156, 159; planning of, 198–200
marketing plan, 117–120
marketing planning: budgets and, 135–136, 138–139; problems in, 149–150
marketing role: financial management questions on, 6; headquarters staff questions on, 6–7; production management questions on, 5–6
market objectives, setting, 120–121

market orientation, 24-28
marketplace: environmental forces on, 18-22; strategy and, 230-237; and technology, 13-18
market research, 51-53; agencies, as sources of secondary data, 61-62
Market Research Corp., 62
market segmentation, 230-234
market share: analysis, 96-97; strategy and, 172
market stabilization and price objective, 212
McGraw-Hill Book Co., 237
Mead Johnson, 27, 35
media, as secondary data sources, 60-61
mid-management authority, 49-50
models: allocation, 135; in decision making, 125-129

National Bureau of Economic Research, 63
natural expense, 99-100
new product: development and market orientation, 25-26; strategies, 200-207
New York Times Index, 60
Nielsen, A. C., 62
nonrecurring decisions, 124-125

objectives, 39-43; price, 211-214
observation, in primary data collection, 71-73
one-time decisions, 124-125
opportunities, objectives in identifying, 40-41

Paley, Norton, 153n
Pemco Co., 183-185
penetration pricing, 215
performance analysis, 93-102
performance objectives, 40
PERT and CPM, 144-149
Philadelphia Quartz Co., 169
Philip Morris Inc., 188-189
planning: audit of, 139-140; benefits of, 34-39; job of, 33-34; of marketing mix, 198-200
policies, 43-45; pricing, 214-220
Polk, R. L. Co., 62
price: determination and administration, 157-158; strategies, 211-214
price-cutting, combating, 219-220
price-lining, 218
pricing policies, 214-220
primary data: defined, 53-54, 55-56; collection methods, 65-75
probability theory, 129-130
product: acceptance and strategy, 172-173; market segmentation by, 232; planning and development, 156-157; strategies, 200-211
product life cycle, 17; promotion and, 224-225; strategy and, 173-177, 208-209
production: management, questions on marketing role, 5-6; orientation, 23-24
profit, by unorthodoxy, 178-179; cost analysis, 98-100; maximization as price objective, 213

program evaluation and review technique (PERT), 144–149
project flowcharts, 143–144
promotion, 158; strategies, 221–226
promotional mix, 222–226
psychographics, market segmentation by, 233
psychological pricing, 218–219

questionnaire, in primary data collection, 67, 68

repetitive decisions, 124
research and product development budget, 139
resources and strategy, 164
responsibilities: assigning, 43; objectives in identifying, 40–41
retail feedback and strategy, 172
return on investment, 102, 103–106; as price objective, 212
risk, planning in reducing, 37–39
role-playing in decision making, 126

sale price vs. cost, 22
sales: analysis, 94–96; strategy and, 171–172
sales and administrative expenses, estimates of, 138–139
sales budget, 138
sales force analysis, 101–102
sample, in primary data collection, 68–69

schedules, 140–141; project flowcharts and diagrams, 143–144; timetables of deadlines, 141–143; PERT and CPM, 144–149
Scott, O. M. and Sons, 226
secondary data, 53–57; external sources, 58–65; internal sources, 57–58
semivariable costs, 115
Shenker Corp., 183–185
short-range marketing plans, 119–120
short-term objectives, 40
simulation, computerized, in decision making, 126–127
single-price policy, 217–218
skim-the-cream pricing, 215
social constraints, strategy and, 163–164
Sony Corp., 179
speed and strategy, 168–177
spending strategy, distribution, 228–230
Stanford Research Institute, 63
strategy: activity zones, 193–197; alternate objectives, 192–193; competition and, 194–196; concentration and, 186–192; controllable factors, 155–159; customer's environment and, 196–197; distribution, 226–230; grand, 166–167; indirect approach, 177–180; marketplace and, 230–237; price, 211–214; product, 200–211; promotion, 221–226; speed and, 168–177; in unbalancing competition, 180–186; uncontrollable

factors, 159–166; working, 167–168
survey, in primary data collection, 66–71
Swift and Co., 52

technology, change and, 13–18
3M Co., 210
Tiffany, 207
timing: in forecasting, 112; guidelines, 171–177
trade associations, as secondary data sources, 62
transportation mode, distribution and, 228–229
Twentieth Century Fund, 63

U.S. Rubber Co., 52
universities, as secondary data sources, 63
University of Chicago National Opinion Research Center, 63
University of Michigan Survey Research Center, 63
utilization of data, 92–93

variable costs, 114–115
varying prices, 218

Wall Street Journal Index, 60
warehouse: distribution, 229–230; feedback, strategy and, 172

Kirtley Library
Columbia College
Columbia, Missouri 65216